Critical Care Patient Transport

Principles and Practice
5th Edition

Richard A. Patterson
MBA, NR/CCEMT-P, MICP, FP-C, CFI, CFI-I, AGI, IGI

Christina D. Patterson
MS, RN, CCRN, CEN, CFRN, CTRN, NREMT-P

Critical Care Patient Transport
Principles & Practice. 5th Edition

Richard A. Patterson
Christina. Patterson

Printed in the United States of America,
Cover Image: ©Bell 222 HEMS Program

Patterson, Richard A.
Patterson, Christina D.

     ISBN-10: 0615242677

     ISBN-13: 978-0615242675

Library of Congress Control Number: 2008909672

To my wife and best friend of fifteen years;

Christina; without you, I could not have accomplished many of the successes that I have. Your love, support, and friendship have helped to give me the strength and confidence that was needed to accomplish the many things that I am proudest of.

To my daughters;

You are my Sunshine and Rain; for without you there would be no life or joy. You have been the true motivation and inspiration in my life and are the best thing that has ever happened to me. Your loving patience has taught me the meaning of fatherhood.

To my Dad;

To my riding buddy, best friend, and Father; I want to thank you for helping to instill values and beliefs that ultimately led me to being the man that I am today. Many times as a young child, I did not understand your passion for honesty, loyalty, and fairness. You demanded the very best from me and accepted no less than 110% effort. Now as a Father myself, I completely understand and thank you for all that you have done for me.

To my Mother;
12/25/50 - 4/10/08

Cancer took you from all of us this year, not allowing me a chance to say Good-Bye to you. Your swift passing has left me with a void that will never be filled. I will always miss your smile, laughter, and passion for life. I hope you find Vienna in the Heavens!

And lastly, special thanks TO YOU for purchasing this textbook and our CD Audio packages, as well as the many fixed-wing and rotor-wing Air Medical programs, Nurses, Paramedics, and other healthcare workers that continue to allow us into their homes to assist with the training of their staff. Without their assistance and help coordinating these courses, Critical Care Concepts, Inc. would not be where it is today.

This *"Critical Care Patient Transport"* **Review Course** has been re-formatted to accommodate persons wishing to review for the "Certified Flight Nurse" (CFRN), "Flight Paramedic - Certified" (FP-C), "Certification in Emergency Nursing" (CEN), as well as the "Certified Transport Registered Nurse" (CTRN) certifications. It has been reviewed thoroughly and approved by the Virginia Department of Health, Office of EMS.

Our CFRN and FP-C Review Course is now an **Approved FP-C Review Course** by the *Board for Critical Care Transport Paramedic Certification* (BCCTPC) and meets the requirements for initial as well as recertification of the Flight Paramedic – Certified (FP-C) certification.

The fill in the blank answers are at the end of the book

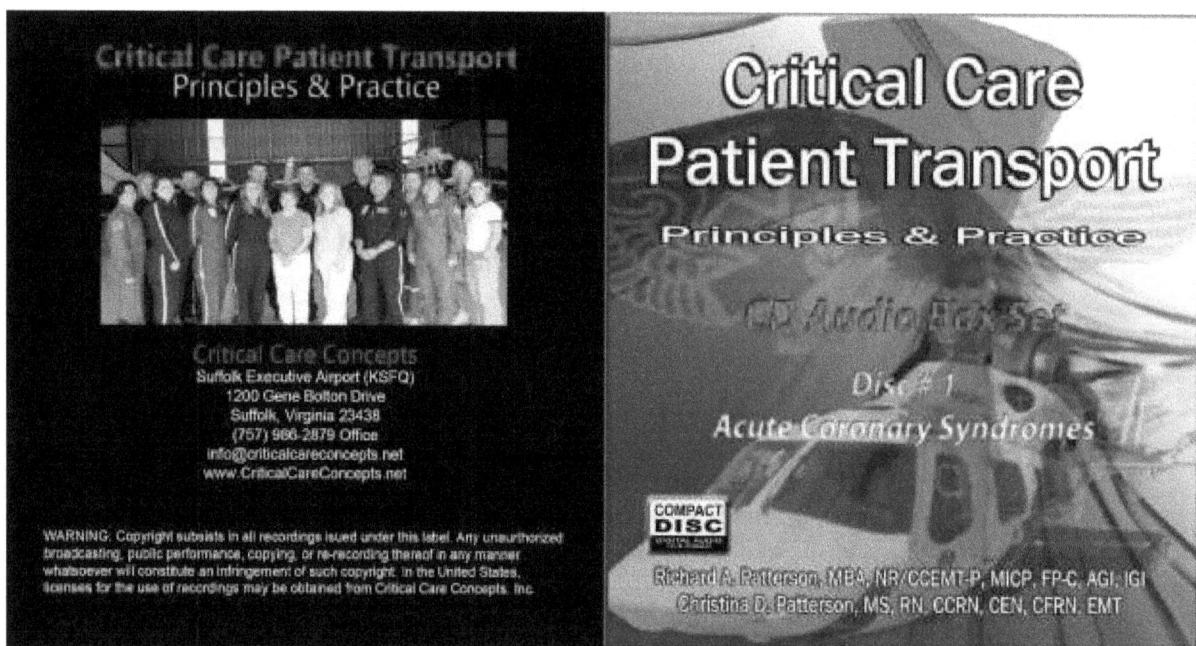

Also available, our **8 Disk CD Audio Box Set** to accompany this textbook. It was digitally recorded and edited during several of our live two (2) day review courses; adding much more insight into many of these difficult subjects, allowing the reader to study at their leisure, while at work, driving to work, or in the comfort of your home.

Table of Contents

3.	Special Thanks
4.	Accreditation of course
6.	Welcome Letter
9.	Authors and Biographies
12.	Course Outline - Day 1
15.	Introduction to CFRN and FP-C Certification
21.	Flight Physiology
23.	Cardiac Emergencies & Heart Failure
37.	Ventilator Management & Respiratory Review
43.	Respiratory Parameters
45.	Aircraft Safety, Survival, and CAMTS
51.	Stressors of Flight
55.	Thoracic Trauma
61.	Abdominal Trauma
69.	Radiological Interpretation
73.	Toxicological Emergencies
77.	RSI / Advanced Airway Procedures
85.	Medical Emergencies
89.	Obstetric Emergencies
93.	Gynecological Emergencies
97.	Cardiopulmonary Hemodynamics
103.	Intra Aortic Balloon Pump management
109.	Acid Base Balance / ABG Interpretations
113.	Burn Management
117.	Neurological Emergencies
127.	Pediatric Emergencies
133.	Neonatal Emergencies
135.	Ocular, Musculoskeletal, Integumentary, Bites & Stings, etc,
147.	Review Questions

This review book is intended as a means to assist the Flight Nurse or Flight Paramedic, and Emergency or Transport Nurse / Paramedic in obtaining certification in the Critical Care Transport or Emergency Department environment. The entirety of this course and all materials are covered under Copyright Laws, and any reproduction is forbidden, unless permission has been granted in writing by the author or it's appointees of Critical Care Concepts, Inc. Redistribution of this document is also strictly forbidden and is prohibited.

This review handbook is designed to facilitate learning and increase knowledge retention while preparing for certification as a Certified Flight Registered Nurse (CFRN), Flight Paramedic – Certified (FP-C), Certified Emergency Nurse (CEN), or as a Certified Transport Registered Nurse (CTRN). It is highly recommended that the entirety of core topics discussed in this handbook, are reviewed thoroughly and adequate research time has been allocated prior to taking any certification exam.

You will notice places throughout this handbook that have spaces left out. This was done intentionally, to allow the reader an opportunity to take notes and to review key concepts that need to be retained. Retaining material and adult learning is accomplished by a multitude of ways; and writing down information will assist you in remembering this and recall it later.

Upon completion of the Flight Nurse and Flight Paramedic Critical Care Transport Review Course, the student will receive a certificate of completion and will be approved for 16.5 contact hours.

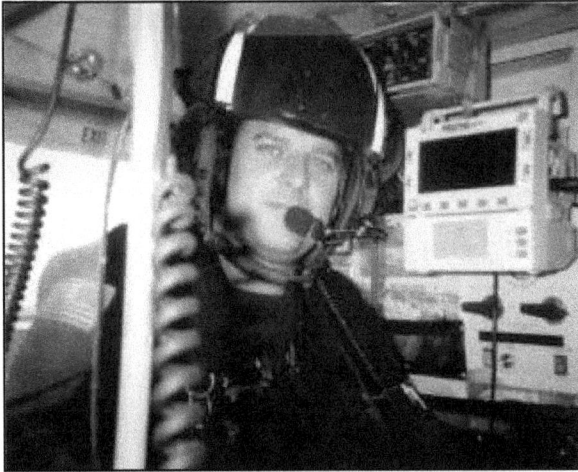

Richard A. Patterson
MBA, NR/CCEMT-P, MICP, FP-C, CFI, CFI-I, AGI, IGI

In addition to being a Certificated Flight Instructor for Airplanes & Helicopters, Richard is also an Instrument rated Commercial Pilot (Multi Engine, Single Engine, and Helicopter), and is certified by the FAA as an Advanced & Instrument Ground Instructor. Richard has experience with the FAA's Regional Safety Team for the Richmond, Virginia FSDO (Flight Standards District Office). He was awarded the prestigious "Master" level safety award from the FAA, and has worked in Emergency Management and as a Flight Paramedic (Both Fixed & Rotorwing) for the last 18 years.

He has managed several air medical programs, has helped start two programs from the ground up, and has also worked as Base Supervisor for several CAMTS accredited rotor-wing programs. He is Instructor credentialed in BCLS, ACLS, PALS, ITLS, NRP, PEPP, PHTLS, ENPC, and EMT-Paramedic. Richard has provided training programs in approximately 15 US States, as well as Canada, South America, Central America, and South Korea, in a multitude of areas. Richard has had his works published in many journal publications, as well as contributing to the current Paramedic textbook; "Paramedic Practice Today", and authored the textbook for air medical providers, "Deciding WEATHER To Fly."

Christina D. Patterson
RN, MS, CCRN, CFRN, CEN, NREMT-P

As the founding CEO for Critical Care Concepts, Christina brings a multitude of knowledge and experience to this textbook and online review course. Currently working as a Director of Critical Care and Emergency Services for a local Trauma Center, Christina is responsible for operations of the Intensive Care units, Critical Care units, Telemetry, step-down units, and the Emergency Department. She is Instructor credentialed in BCLS, ACLS, PALS, NRP, ENPC, TNCC, and is the Regional Ambassador for the NC area Chapter of AACN.

Her work has been published nationally in various critical care nursing journals, and she received the coveted national award from "Who's Who in Business", and "Best of the Best" in Female Executives. She has worked in Critical Care and Emergency Departments with various Level I & II Trauma Centers in Pennsylvania, New York, North Carolina, and Virginia for over the last decade.

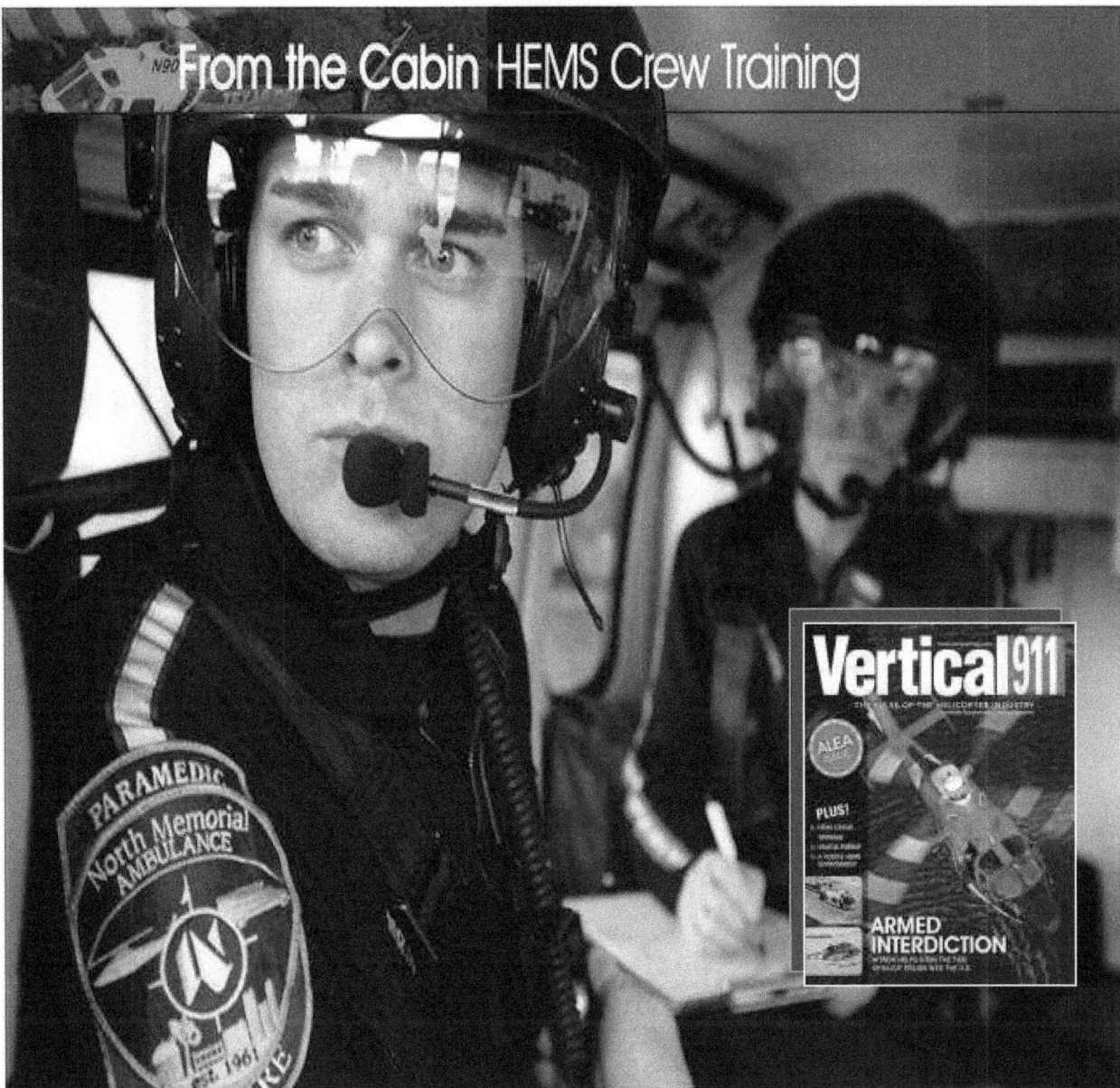

Training Required

BECOMING A FLIGHT PARAMEDIC AND/OR
FLIGHT NURSE IS A LONG AND ARDUOUS
PROCESS THAT REQUIRES AN EXTRA LEVEL OF
DEDICATION AND DETERMINATION.

Story by Richard A. Patterson | Photos by Mark Mennie

For many registered nurses (RNs) and paramedics, a common goal is to work in an environment where critical thinking, autonomy, teamwork and the challenge of positive patient outcomes in complex care situations occurs on a regular basis. These health care professionals welcome the opportunity to rapidly determine the involvement and extent of a patient's injuries. And, when on the emergency transport side, they welcome the opportunity to assess and stabilize critically injured trauma patients and high-acuity inter-facility patients, while determining and utilizing the swiftest means of conveying them.

Decreasing the time spent out of a hospital environment increases a trauma patient's chances of survival (a basic principle of the "golden hour" concept). This is where many of the Title 14 Code of Federal Regulations Part 91 and 135 helicopter emergency medical services (HEMS) and fixed-wing air ambulance transport services come into play. Air medical transport's primary purpose has always been to decrease the time that trauma patients spend between those facilities that do not have the capabilities to treat their injuries or illnesses, and those that offer a higher level of care (typically Level I or Level II trauma centers).

Over the years, though, in addition to decreasing transport time, air medical services now generally offer a higher scope of care and formulary. Also, air medical crews now normally have a more advanced skill set and higher qualifications than their facility counterparts. For instance, while RN training programs usually last two years (some university programs are four years), and the same goes for paramedic training, flight nurses and flight paramedics train well beyond those initial certifications.

TRAINING, TRAINING... AND MORE TRAINING

So, exactly what goes into becoming a flight nurse or flight paramedic? Once HEMS crew candidates complete their initial nurse or paramedic training, they need real-life experience managing and treating complex medical and trauma patients. For this, they need a minimum of five years in a busy EMS environment or in a critical care/intensive care unit or emergency department, before an air medical program will even entertain their resume.

A newly licensed RN will usually go to a medical/surgical unit or step-down telemetry unit. There, he or she will be responsible for taking care of as many as 15 or 20 patients at a time — but generally patients who are in very stable conditions. From the medical/surgical unit, an RN can move into a more demanding role as an intensive care nurse or emergency room nurse.

Newly certified paramedics, meanwhile, follow a similar career path, usually starting by transporting stable dialysis or medical appointment patients. From there, they can move to more demanding roles as 911 providers in busy rural or urban environments, thus getting exposure to victims of drug overdoses, shootings, stabbings and the like.

So, now the paramedic or RN has the five years in a busy working environment... but they must show a desire to excel even further. This is usually accomplished by teaching many of the related certifications that come with being an RN or paramedic (e.g., Advanced Cardiac Life Support, Pediatric Advanced Life Support, Basic Cardiac Life Support, Trauma Nursing Core Course and International Trauma Life Support certifications). When a health care

professional starts training fellow workers, physicians and EMS personnel for these certifications, they begin to hone their skill set and master complex issues and treatment options — making them better caregivers and providers. And, as instructors, they make themselves more marketable to prospective employers, who can save a few dollars by keeping their training in-house.

Having dual credentials is another way to stand out. As a former manager at various air medical programs throughout the United States, I would always be impressed if an applicant were not only an RN or nationally registered paramedic, but were both. From my standpoint as a manager, this meant I could use them in either role, filling scheduling needs as necessary (as long as they currently met the requirements for both roles, of course). I knew these dual-credentialed health care providers would be better clinicians, as they knew and would understand the other providers' roles and responsibilities. Dual credentialing also showed me they had a strong desire and dedication to their chosen professions and the outcome of their patients. And, it added to their experience level, giving them exposure to a multitude of patient conditions that they likely would not have had otherwise.

OPPOSITE Flight paramedics and flight nurses are at the tops of their retrospective professions. Here, we see a flight crew with North Memorial Air Care in the Minneapolis, Minn., area.

BELOW (ALL) Air medical crews at work.
BOTTOM LEFT **Claes Axstål Photo**

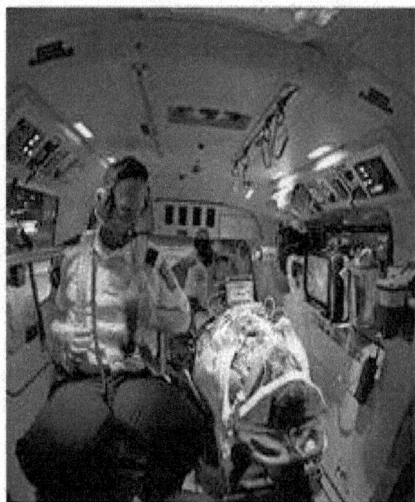

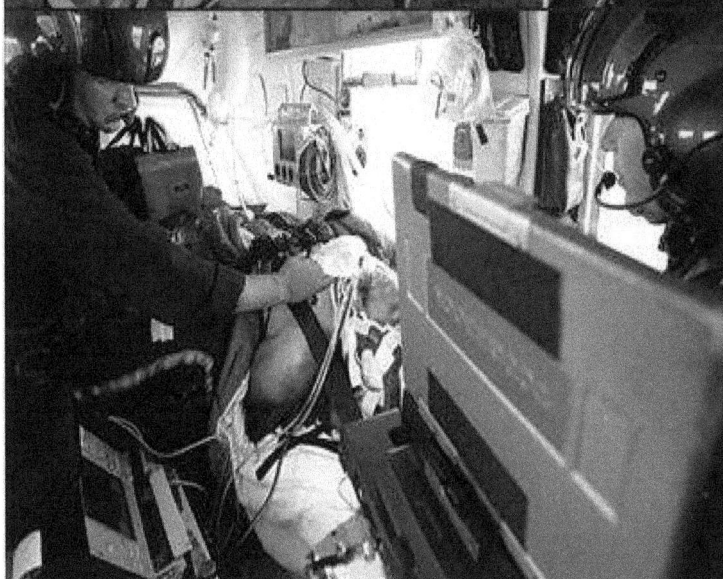

Flight nurses and flight paramedics must be able to stay focused on the mission, even in the midst of stressful and emotionally demanding circumstances.

ADDITIONAL CREDENTIALS

Beyond the typical five years experience, the variety of instructor certifications and becoming dual-credentialed, there are also specialty certifications directly relating to the critical care and transport environments. The main ones are the Flight Paramedic - Certified (FP-C) and the Certified Flight Registered Nurse (CFRN) qualifications. These may be two of the most demanding credentials that a nurse or paramedic can prepare for and obtain.

Some other newer credentials are also available, including Certified Transport Registered Nurse (CTRN), and the newest, Critical Care Paramedic - Certified (CCP-C). The industry standard, however, has always been that an applicant needed to obtain the CFRN or FP-C within one year of hire for a ground or air medical transport program.

Currently, there are only about 11 "approved" review course programs in the world that can prepare an applicant for the FP-C or CFRN credentials. (For more information on approved review course curriculum providers, go to www.flightparamedic.org/reviews.htm.) An applicant does not have to take an approved review course, but this will offer the best chance of ensuring that he or she has the correct information to successfully pass these exams: for initial test-takers who did not take a review course, these exams boast a sad 40-percent pass rate nationally.

The topics included on these certification examinations are: flight safety and survival; Federal Aviation Regulations Part 91 and 135 rules and regulations; cardiac emergencies; neurological emergencies; medical and critical care patient management; pediatric and neonatal patient care; burn patient issues; obstetric and gynecological emergencies; toxicological emergencies; radiology interpretation (X-ray); ventilator management and advanced airway skills; and pharmacology. The exams also cover stressors of flight, gas laws and their applications on patient care, and various types of hypoxia.

A WILL TO SERVE

Of course, even before you get to all the training and experience gathering, you have to take a look at whether you have the right personality for these positions. So,

what type of individual would have a desire to achieve all these aforementioned qualifications? Such a person would have to be a goal-driven, career-minded individual — one who desires excellence, and who is usually determined to achieve results. We are talking about a person whose ultimate goal is to decrease the morbidity and mortality of their patients, increase survival rates and improve their patients' quality of life. Only a very strong-willed individual could carry such a burden on himself or herself.

With the multitude of deaths, injuries and psychological issues that a flight nurse or flight paramedic is exposed to, they also have to be able to stay focused on the mission at hand — despite the child or infant that perishes in their arms; the mother who passes away during a prolonged extrication after an automobile accident; or the young man killed on his motorcycle while enjoying the warmth of the sun on his face. All that we as health care providers can do is strive to prevent such misfortunes in the future by educating the public and educating ourselves to the highest level achievable. (And we should never forget to kiss our family members when we leave the house — as it could be us on the receiving end of treatment one day.)

So, you can see why being a flight paramedic or flight nurse is an achievement that illustrates a desire to be the best at what you do. It sets the benchmark of excellence for a nurse or paramedic to achieve, and provides the highest level of care one can offer to patients. The clinicians' dedication to the profession is evident by the multitude of certifications and credentials they have obtained, their continuing quest to be the best in their field, and the oftentimes self-limiting rewards that come with helping your fellow human beings.

Richard Patterson is a certified flight instructor for airplanes and helicopters, and a commercial pilot with instrument, multi-engine and helicopter ratings. He has worked in emergency management and as a flight paramedic for 18 years, and has managed several air med programs. He is head of Critical Care Concepts, a Virginia-approved Advanced Life Support training agency.

Critical Care Transport Review Course

Day 1

7:30 am	Registration Begins
8:00 am	Welcome and Introductions
8:15 am	Aviation Physiology
9:00 am	Cardiac Emergencies
10:00 am	Ventilator Management
11:00 am	Aircraft Safety, Survival, CAMTS
12:00 pm	Working Lunch
12:30 pm	Thoracic Trauma
1:30 pm	Radiology Rounds
1:45 pm	Abdominal Injuries
2:15 pm	Toxicological Emergencies
2:30 pm	RSI / Advanced Airway
3:30 pm	Medical Emergencies Critical Care Patient Management
4:30 pm	Obstetric Emergencies Gynecological Emergencies
5:30 pm	Adjourn for the day

Critical Care Transport Review Course

Day 2

8:00 am	Hemodynamic Monitoring Principles and Practice
8:30 am	IABP / VAD Devices Troubleshooting issues
9:15 am	Acid Base Disorders ABG Interpretations
10:30 am	Burn Management
11:30 am	Neurological Emergencies
12:30 pm	Working Lunch
1:00 pm	Pediatric Emergencies
2:00 pm	Neonatal Emergencies
3:00 pm	Ocular Injuries, Environmental Emergencies Bites and Stings
4:30 pm	CEN, CFRN, CTRN, & FP-C Exam Review Jeopardy Game Review

Introduction from ENA and BCEN on the Exam process

All the latest information regarding the *CEN, CCRN, CTRN, CFRN,* and *FP-C* exam

Beginnings

First CEN and CFRN exam was in 1980
It was administered by paper and pencil with 250 items
It was offered two times per year in approximately 80 locations
Results were mailed in 6 weeks

Then

Paper and pencil became computerized adaptive in 1998
The exam was 100 - 150 items
Because of adaptive format, the candidate could not go back and this was not well received
The exam was offered weekly
Results were mailed in four weeks

Now

The exam is offered five days/week (Monday - Friday)
Exam is offered 2/day at 9:00 am & 1:00 pm
Score reports are obtained before leaving the testing site
The exam is 175 items **AND**
You Can Go Back!!
Items can be reviewed or skipped and returned to before exiting the exam

Test Sites

There is at least one site per state
And more than 125 sites

About the Sites

Exams are taken at AMP assessment centers of which most are at H&R Block offices

Requirements

Current <u>unrestricted nursing license</u> (RN)

Emergency nursing and or flight nursing background is **NOT** required, **but two years experience is recommended**

No BSN Required

Although BCEN believes in and supports furthering one's education, a BSN is not required to take the CEN or CFRN exam

For the FP-C Exam, a Driver's license or *other picture ID is required*

A copy of your Paramedic certification is required

Application Process

Obtain, complete, and mail in an application
The application is processed
A verification letter and handbook are sent to the candidate.
The applicant calls the 800 number and schedules an exam site and time.
The exam is scheduled and taken within the 90 day window.

At The Site

Arrive on time at the site (wait for proctor to arrive)
Present valid government issued identification and sign in
Take your own digital photograph which is displayed during testing in the upper right hand corner

The Exam

After a tutorial, the exam of 175 questions is started
Three hours are allowed to take the exam AND
You CAN skip questions and return to them later

Score Report

Once you have completed the exam, you will receive your score report
If successful, a wallet card and certificate will be sent in 4-6 weeks

CEN Renewal Option
(CEN-RO)

On-line renewal registration now available!
Renew every 4 years using one of the following options:

Option 1:
      CEN-RO by Computer Exam at an AMP Testing Centers.
Apply on-line, no need to request an application

Option 2:   CEN-RO by Continuing Education (CE)
100 Contact Hours required
Submit CEUs directly on-line

Option 3:   CEN-RO by IBT (Internet Based Testing)
Take this renewal test on-line

CFRN Renewal Option (Air-RO)

Renew every 4 years using one of the following options:

Option 1:   Air-RO by Computer   Exam
AMP Testing Centers

Option 2:   Air-RO by Continuing Education (CE)
100 Contact Hours required

For more information:

Call BCEN at (800) 900-9659, ext. 2630
Or E-mail BCEN at bcen@ena.org
Or visit the website at www.ena.org/bcen

Introduction to CFRN and FP-C

FP-C Test Breakdown
- 125 questions (10 Beta test questions)
- Through Board for Critical Care Transport Paramedic Certification (BCCTPC)
- The questions are "Weighted", meaning some questions have different values placed upon them.
- Dependent upon the version of the exam that is to be given at the site.
- 31 recall, 54 application, 40 analysis

FP-C Topics covered

Trauma Management	09
Aircraft fundamentals / Safety	12
Flight Physiology	10
Advanced Airway Mgt.	05
Neurological Emergencies	10
Critical Care Emergencies	20
Respiratory Patient	10
Toxic Exposures	06
Obstetrical Emergencies	04
Neonatology	04
Pediatrics	10
Burn Management Emergencies	16
Environmental Issues	04

CFRN Test Breakdown

- 150 questions
- Through ENA at Prometrix Ctrs nationally
- 3 hours to complete
- 110 correct needed to pass

CFRN Test Breakdown

Cardiopulmonary	27
GI/GU and OB Emergencies	15
Maxillofacial and Orthopedic	07
Neurological Emergencies	14
Multisystem Emergencies	30
Patient Management	37
Safety Issues	12
Professional Issues	08

Recommended Reading

Critical Care Transport, Principles and Practice
Patterson, Richard; 5th^d Ed.,
Critical Care Concepts, 2008

Air and Surface Patient Transport, Principles and Practice
Holleran, Renee; 3rd Ed., ASTNA, Mosby, 2003

Core Curriculum for Critical Care Nursing
Alspach, Jo Anne;, 6th Ed., AANC, Saunders, 1998

Mission Profile
- Air Medical Utilization
- Time - Distance - Environmental Issues
- Acuity
- Scene
- ICU
- E.D.
- Cath lab
- Neonatal

CAMTS Crew Configuration
- EMT-P/RN
- RN/RN
- RN/MD
- Other

Configurations
- Specialty Teams
- Neonatal
- Pediatric
- Health Fairs
- Aircraft and Crew
- Safety In-services
- ECMO
- IABP
- Organ

- Escort

Breakables
- Medical Equipment
- Soft packs
- Reduced weight\size
- Monitors
- No breakables

Mission Profile
- Medical Command
- Typically open protocols
- Limited need for on-line medical contact
- Typically more advanced protocols
- RSI, Central lines, etc.

Mission Profile
- Service Oriented
- Military
- Critical Care
- ALS\BLS
- Search and rescue
- Law enforcement
- Fire Department

Crew Resource Management
Defined as Using ALL Available resources…
including the Medical crew!

- Teamwork - expanded role as aircraft crew
- Navigation
- Weather minimums (IFR/VFR)
- Aircraft avoidance (clock system)
- Public relations
- Emergencies
- Customer Oriented
- Aviation sectional map
- Loran/GPS

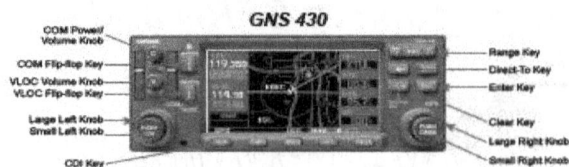

Weather Minimums
- IFR/VFR
- Public Relations

LZ Preparation
- Finding the LZ
- Obstacle avoidance on entrance/exit of LZ
- Working around the Aircraft

Crew Resource Management
- Tact in the trenches
- Public relations with EMS/Hospital providers
- Communicating under high stress environment
- Critical patients under high noise environment
- Communicating effectively with people under the rotor system
- "Bed to bed with chart in hand"
- With everyone's expectations exceeded

Professional Organizations

AAMS	Association of Air Medical Services
AMPA	Air Medical Physician Association
NEMSPA	National EMS Pilots Association
ASTNA	Air & Surface Transport Nurses Association
IFPA	International Association of Flight Paramedics
NAACS	National Association of Air Medical Communication Specialists

Formed 1991
Organized forum for improving services
- Mechanism for voluntary compliance standards
- Marker of excellence to guide federal and state agencies
- Accreditation standards as a benchmark for quality
- Maintain currency for industry standardization

Air Medical Appropriateness
- Who will benefit from air transport?

Indications
- Any adult, child, or neonate
- Acute medical and surgical problem
- In need of specialty care
- At the scene of illness or injury
- Where the time required to transport the
- patient by ground to definitive care is considered to be excessive

Who will not benefit from air transport?

Contraindications
- Patients with terminal illnesses
- Patients with DNR orders
- Patients in cardiopulmonary arrest
- Without return of a spontaneous pulse
- Patients who will overwhelm the crew

Air Medical Efficiency

- Limitations of air medical transport

Aircraft by nature..........
- Are crowded and claustrophobic
- Are noisy
- Compromise performance of CPR
- Contain vibration/movement/poor lighting
- Limit the senses of the care provider
- Are prone to extreme temperatures

Air Medical Efficacy

- Are there measurable benefits of air transport?
- Intuitively - makes sense
- Decreasing the transport time
- Faster means of transportation
- More direct route of travel
- Quantitative Data - unavailable
- Precise time savings
- Increased level of care
- Effects on morbidity and mortality
- "An aircraft is just one tool in the medical transport toolbox..
- ...... it's all in how it gets applied!

Aircraft Orientation (FW)

1. Components of aircraft construction
2. Concepts of aerodynamics
3. Weight and Balance
4. Ground effect
5. Statute vs. nautical miles
6. Airport ramp dangers
7. Safety requirements

Aircraft components

Fuselage
- Flight deck
- Cockpit
Wings
- Fuel storage
- Engine installation
- Propeller

Empennage
- Vertical fin
- Horizontal stabilizer

Lights
- Starboard (right)
- Port (Left)
- Anti-collision
- rotating beacon
- Strobes

Airflow over wings / Angle of Attack

Lift

Lower Pressure

Faster Airflow

Slower Airflow

Higher Pressure

4 Forces of Flight

- Lift
- Thrust
- Weight
- Drag

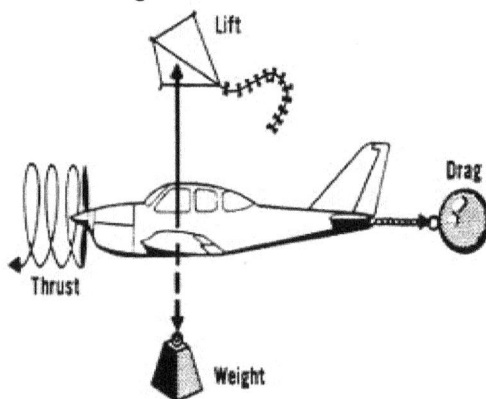

Lift

Drag

Thrust

Weight

Axis of Control
    Roll
    Pitch
    Yaw

Flaps
    Increase descent rate w/o
    increasing airspeed

Center of Gravity (CG)
    Reference point for weight and
    balance calculations
    Loading considerations
    Max gross weight
    Conditions affecting gross weight
    This is why we have weight limits
    for crew!

Weight and Balance
    Maximum weight

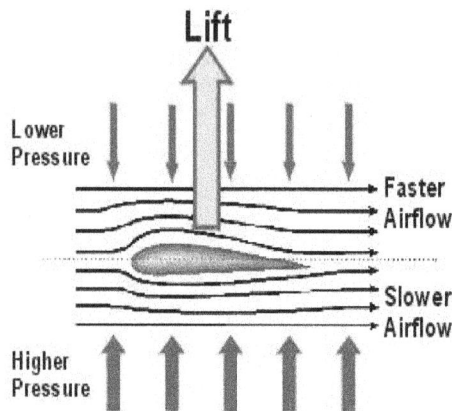

Distribution of weight
    Aft / forward loading

Angle of Attack
    Pilot controls
    Chord line
    Importance in landing

Ground effect
    Boundary layer between aircraft
    and ground
    Aircraft performs more effectively
    in GE
    Augments landing procedures

Measurements
    Statute Mile
    Nautical Mile
    Knot

Weather
Pressure systems
    High pressure ridges
    Low pressure troughs
    Fronts

Dew Point
    Vapor Information
    Fog when narrow dew point
    spread

Frontal condition weather
    Warm fronts
    Cold fronts

Cloud types
    Cumulus - Cumulonimbus
    Altostratus - altocumulus
    Cirrus – Cirrocumulus

Thunderstorms

Characteristics
    Winds
    Hail
    Heavy rain
    Wind shear
    Reduced visibility
    Turbulence
    Warning signs

Altitude related complications, physical gas laws, changes in atmospheric pressure, temperature, and volume

8 stressors of flight

Common environmental phenomenon
Not all stress is altitude related

Principles of Atmospheric pressure

At sea level, the weight of a one square inch column of air that extends to the outer limits of space is termed "one atmosphere" (ATM)
1 ATM weighs 14.7 lbs (760mm Hg [torr])

AGL (Above Ground Level)
MSL (Mean Sea Level)

Flight Physiology

The earth's atmosphere at sea level is
Made up of a collection of gases that

Nitrogen	78%	593 mmHg
Oxygen	21%	160 mmHg
Other	1%	7 mmHg

Barometric pressure:

Principles of Atmospheric pressure
As you ascend, the pressure becomes less
(0.5ATM or 380mm Hg at 18,000 ft)

As you dive in water, you increase the forces (or weight) on your body by 1ATM for every 33 feet you are submerged. Why we call it diving "1 atmosphere."

Flight Physiology

1. Physiologic Zone
2. Physiologically deficient zone
3. Space-equivalent zone
4. Space

Gas Laws & their applications

Boyle's Law (expansion)

"At a constant temperature, the volume is inversely proportional to the pressure"
$(P1)(V1) = (P2)(V2)$

As altitude increases, the gas

_____

As altitude decreases, the gas

_____

Boyle's Law *(Balloon)*

Effects on patients

a.

b. Gastrointestinal changes
Wet gases expand more than dry
Normally, 1 liter of gas remains in the GI tract and can expand 1-1.5 times original volume

c.

d.

e.

Effects on equipment

a.

b.

c.

Charles' Law (volume & Temperature)
"Charles is a hot-head"

"When the pressure remains constant, the volume is nearly proportional to its absolute temperature." i.e.: A can of hair spray in the fire, or filling O2 cylinders
$V1/T1=V2/T2$

Effects on patients

Gas volume expands as temperature _____

Gas volume shrinks as temperature _____

Climb 1000 ft = 2°C drop

Gay-Lussac's Law
"Charles's gay brother"

Directly proportional relationship between

Temperature and _____.

$P_1/T_1 = P_2/T_2$

Similar to Charles Law

Henry's Law (solubility) *"Heineken"*

"The amount of gas dissolved in 1 cm3 of a liquid is proportional to the partial pressure of the gas in contact with the liquid"

"The partial pressure of a gas and the solubility of the gas determine the amount of gas that will dissolve into a liquid"

"Gases have a tendency to move from a higher concentration to that of a lower concentration"

Graham's Law

"The diffusion rate of a gas through a liquid medium is directly related to the solubility of the gas & inversely proportional to the square root of its density."

Means gases diffuse from a higher concentration to an area of lower concentration.

Dalton's Law (partial pressures)
"Dalton's Gang"

"The total pressure of a gas mixture is the sum of the individual or partial pressures of all the gases in the mixture" (*"Dalton's Gang)*

$P_t = P_1 + P_2 + P_3 + P_4...$

Gas Laws review...

Boyle's Law can affect any body cavity or piece of equipment that has an enclosed air space

Dalton's Law can affect the oxygen transfer into the bloodstream

Henry's Law can affect evolved gases in the bloodstream; a rapid loss of pressurization can cause nitrogen gas bubble formation. *(Heineken)*

Crew ability to manage affects of gas laws

Cabin
- Pressurized Cabin adjustable
- artificial rapid decompression

Unpressurized
- Fluctuations in air pressure

Body cavities
- ETT cuffs
- PASG
- IV fluids

Stressors of Flight
- Environment
- 8 Stressors of Flight
- Hypoxia
- Barometric pressure
- Thermal changes
- Decreased humidity
- Noise
- Vibration
- Fatigue - Base metabolic rate
- G Forces

Factors effecting flight stressors

D -

E -

A -

T -

H -

PAO2? 5mm Hg per every 1000 foot Δ in altitude

Oxygen Adjustment calculation:

(%IO2 x P1) = %IO2 for altitude
P2

O2 Adjustment example

You are called to rendezvous with local EMS to transport a 5 year old male patient with Epiglottitis. The patient is currently on humidified oxygen at .40 FiO2. The EMS service is at an altitude of 3,000 feet (681 mmHg.) You have a maximum altitude of 8,000 feet (565 mmHg.). What percentage of oxygen will you have to increase the patient to maintain his oxygen saturation during flight?

Cardiovascular Emergencies

Objectives
1. Review normal cardiac anatomy and physiology.
2. Explain the coronary circulation structures and the importance of perfusing the heart.
3. Demonstrate an understanding of cardiovascular disease and management techniques.
4. Differentiate pharmacologic management of cardiac disease processes.
5. Identify the types and management of various MI's Anatomy and Physiology

Anatomy and Physiology of the Heart

Pericardium

Heart
- Fibrous Skeleton
- Epicardium – thin outermost layer
- Myocardium – thick muscular middle layer
- Endocardium – thin inner layer
- Muscle
  - Striated
  - Intercalated cells allow two synctiums
  - Inner and Outer spiral, middle circular

Heart Chambers
- Four chambers divided by a septum
- Two types: Atrioventriculur and Semilunar
- AV Valves: located between Atria and Ventricles

- Semilunar Valves: located between left and right halves of the heart

Valves Open as a result of <u>increased ventricular pressures</u> and close as a result of <u>decreased ventricle pressures</u>

Valve order as blood flows: "Toilet Paper My Ass"
1. Deoxygenated blood enters the RA via the Inferior and Superior Vena cava.
2. Blood then travels from the RA to the RV via the "Tricuspid" valve.
3. From the RV via the "Pulmonic" valve through the PULMONARY ARTERY, which happens to be the only artery that carried unoxygenated blood. We will discuss Pulmonary Artery Catheters and monitoring later in Hemodynamics.
4. It then goes to the lungs to pick up O2, and offload CO2.
5. Oxygenated blood then leaves the lungs via the Pulmonary Vein (Which happens to be the only vein carrying oxygenated blood) to the LA.
6. It then goes from the LA via the "Mitral" or Bicuspid (As it has 2 flaps) to the LV, where it is pumped systemically through the Aortic valve, Aortic Arch to the body.

Fun with Formulas (You won't have to calculate them!)

Parameter	Method Calculation	Normal Values
Mean Arterial Pressure (MAP)	$\frac{(Diastolic \times 2) + Systolic}{3}$	70-105 mmHg.
Cardiac Output (CO)	Liters per minute	4-8L
Cardiac Index (CI)	CO / Body Surface (BSA)	2.5-4.0 L/min/m2 2.2 with Cardiac History
Stroke Volume (SV)	CO X HR	50-100 ml/beat
Stroke Index (SI)	SV / BSA	25-45 ml/m2/beat
Systemic Vascular Resistance (SVR)	$\frac{(MAP - CVP) \times 80}{CO}$	800-1200 dynes
Pulmonary Vascular Resistance (PVR)	$\frac{(Mean\ PA - PCWP) \times 80}{CO}$	50-250
Left Ventricular Stroke Work Index (LVSWI)	SI x (MAP – PCWP) x 0.0136	40-65 g/m/m2
Right Ventricular Stroke Work Index (RVSWI)	SI x (Pam-CVP) x 0.0136	5-12 g/m/m2
Right Atrial Pressure (RAP)		2-6 mmHg.
Central Venous Pressure (CVP)		2-6 mmHg.
Right Ventricular End-Diastolic Volume (RVEDV)	Measured with REF cath	100-160 ml.
Pulmonary Artery Pressure (PA)	MPAP = 9-18	PAS 15-30 PAD 5-15
Pulmonary Artery Wedge Pressure (PCWP)	Pulmonary Artery Occlusive Pressure (PAOP)	4-12
Coronary Artery Perfusion Pressure (CAPP)	CAPP = Diastolic - PCWP	60-80 mmHg.
Right Ventricular Ejection Pressure (REF)	REF cath	40-60 %
Ejection Fraction (EF)		55-75
Arterial Oxygen Saturation (SaO2)	Measured by ABG	95-100 %
Venous Oxygen Saturation (SvO2)	Measured by mixed venous gas or SvO2	60-80 %
Arterial Oxygen Content (CaO2)	1.34 x Hgb x SaO2	18-20 ml / dl
Venous Oxygen Content (CvO2)	1.34 x Hgb x SvO2	12-16 ml / dl
Oxygen Delivery (DO2)	CO x CaO2 x 10	900-1100 ml / min
Oxygen Consumption (VO2)	CO x Hgb x 1.34 x (SaO2 – SvO2)	200-300 ml / min
Oxygen Extraction Ratio (OxER)	$\frac{CaO2 - CvO2}{CaO2}$	22-30 %

Ventilator Management

"Mechanical ventilatory support is applied in order to provide or maintain ventilation and carbon dioxide removal from the lungs and to provide adequate arterial oxygenation to aid in oxygen delivery to the tissues."

Indication for ventilation:
- Ventilatory Impairment
  - evidenced by: Alveolar hypoventilation

Ineffective minute volumes
- Hypercapnia
- Academia
- Hypoxemia.

Shunt hypoxia
- Ineffective uptake of O2 despite increased levels of delivered oxygen due to a physiologic shunt
- Hyperventilation
- Hypoxemia
- Hypocapnia
- Alkalemia
- Bronchiolitis
- Pneumonia
- pulmonary edema
- ARDS
- loss of surfactant

Many patients will have both indications concurrently

Ventilatory Care Goals
- Provide mechanical power to maintain physiologic ventilation
- Manipulate ventilatory pattern and airway pressures
- Decrease work of breathing (WOB)
- Decrease myocardial work

Indications
- Apnea

- Acute respiratory failure
- Impending ventilatory failure
- Hypoxemia

Principles of Ventilatory Support
- Oxygenation
- PO2
- Ventilation
- PCO2
- pH

Types of Ventilation

Pressure cycled ventilation
- Inspiration ends at a pre-set airway pressure
- Volume per breath may be variable
- Lungs are relatively free of resistance and are compliant
- Patient is unconscious, sedated or cooperative
- Usually support needed for less than 24 hours
- May be used for long-term chronic support

Volume/time cycled ventilation
- Most popular and easily applied
- Essential parameter to control is volume delivery
- Tidal volume (VT) and minute volume (VE) are predictable

Modes of Ventilation

1. Control
2. Assist/Control
3. Synchronized Intermittent Mandatory Ventilation (SIMV)
4. Continuous Positive Airway Pressure (CPAP)
5. Pressure Control
6. Pressure Support

Control
- All parameters of the ventilator cycle (frequency, VT, flow rate) are controlled by the ventilator
- Patient is "locked out" from triggering a breath

- Patient has no active role in ventilatory cycle
- Not routinely used

Assist/Control
- Inspiratory effort greater than sensitivity setting can trigger additional breaths
- Flow rate and VT are controlled

Benefits
- Controls ventilatory pattern

Cautions
- May "stack" breaths and lead to air trapping
- Adjust sensitivity

SIMV
- Ventilator delivers machine breaths
- Base respiratory rate and VT are set
- Synchronized with spontaneous breaths
- Spontaneous breaths possible through circuit
- Spontaneous breaths do not trigger a ventilator cycle
- Flow rate and VT are patient controlled
- Keeps respiratory muscles active and coordinated
- Decreases barotraumas and need for relaxants

CPAP
- Spontaneous breaths over an elevated baseline pressure
- VT and flow are completely patient controlled
- Must set back-up ventilation parameters

Benefits
- Increases compliance and decreases Atelectasis
- In some cases
  - Increases PaO2
  - Decreases work of breathing (WOB)

Pressure Control
- Regulated during inspiration, so VT delivered within a certain pressure limit
- Calculates flow rate so pressure maximum is not exceeded (PRVC)
- Difficult if patient is awake

- VT not assured: monitor closely for hypoventilation

Pressure Support

- Pressure set to enhance gas flow during a spontaneous breath
- Overcomes airway resistance caused by ETT
- Frequently used for weaning

Terminology
- Fraction of Inspired Oxygen (FiO2)
- Tidal Volume (VT)
- Deadspace (VD)
- Frequency (f)
- Minute Ventilation (VE)
- Flow Rate
- Inspiratory Time
- I:E Ratio
- Airway Pressure
- Actual
- Mean
- Peak
- Compliance
- PEEP
- Positive End Expiratory Pressure)/CPAP

Fraction of Inspired Oxygen
- Oxygen concentration, expressed as fraction in decimal form
- e.g. 50% O2 = FiO2 0.5

Tidal Volume (VT)
- Amount of gas moved in one normal breath, expressed in ml
- Normally about 500-600 ml for spontaneously breathing resting adults
- Ventilator 6 -10 cc/Kg

Deadspace (VD)
- Volume of VT gas that remains in the upper airways and does not participate in alveolar gas exchange
- Normally about 1/3 of VT (VD/VT Ratio)
- Extending ventilator circuit improperly may increase deadspace

Frequency (f)
- Breaths per minute

Minute Ventilation (VE)
- VT x frequency
- Relates directly to PCO2 (varies inversely)
- Blood gas analysis is required to determine therapeutic VE

Flow Rate
- Inspiratory (I) flow measured in lpm
- Maintain desired I : E ratio
- Flow may affect pressures

In Time Cycled Ventilation:
- flow rate x I time = VT
- 30 lpm x 1.5 seconds = 750 ml
- 0.5 lpm x 1.5 seconds = 0.750 L

Inspiratory (I) Time
- Amount of time to deliver a single breath, measured in seconds

In Time Cycled Ventilation:
- I time x flow rate = VT
- If frequency is 12 seconds, and I time is 1.5 seconds, what is the Expiratory (E) time?
- Each breath cycle is allowed 5 seconds

I: E Ratio
- Ratio of time for I:E is normally 1:2
- Clinical situations may require ratio to change
- Inverse ratio ventilation
- Delivers breaths with I time greater than E time
- Used for patients with ARDS
- Some ventilators will not allow for inverse I:E ratios

Airway Pressure
- Actual (Paw)
- Real-time airway pressure
- Mean (MAP)
- Mean pressure over one complete ventilatory cycle
- Peak (PIP)
- Highest pressure over a single ventilatory cycle

Compliance

- Measure of the resistance of the lungs to a positive pressure breath
- Increased compliance
- Lungs are more receptive to a mechanical breath
- Reflected in lower airway pressures
- Decreased compliance
- Lungs are less receptive to a mechanical breath, and airway pressures increase
- Patient needs to be suctioned or has bronchospasm
- Restrictive or obstructive lung disease or process
- Developing ARDS or pulmonary edema

PEEP & CPAP

PEEP
- Creates positive end expiratory pressure (above normal atmospheric pressure) to allow alveoli to remain open, thereby enhancing gas exchange at the alveolar level
- Allows for lower FiO2
- Increased FRC
- Improved oxygenation
- Increased compliance in some cases
- Decreased WOB and Atelectasis

CPAP
- PEEP on a spontaneously breathing patient

Clinical Guidelines
- Tidal volume (VT)
- Respiratory rate (f)
- FiO2
- Flow rate
- PIP
- Minute volume (VE)
- Sigh
- Sensitivity

High Pressure Limit

Low Pressure Limit
- 10-15 ml/Kg
- 10-20 breaths per minute

ABG (PO2) or SpO2
- 30-60 lpm (I:E ratio)

- < 40 lpm

ABG (PCO2/pH) ETCO2
- 1.5-2 times VT
- -2 cm, adjust as tolerated
- 10-15 cm above PIP
- 10-15 cm below PIP

Ventilator Controls

1. Mode

2. Control

3. Assist Control

4. SIMV/CPAP

5. Cal

I Time / VT
- 0.1-3.0 seconds
- 50-2000 ml
- Default is volume cycled

Breath Rate
- 0-150 per minute

Flow
- 5-100 lpm

Assist Sensitivity
- -2 to -8 cm H_2O

PEEP compensated
- Manual PEEP Reference
- PEEP is read by ventilator
- 0-20 cm H_2O

Pressure Relief
- Sets maximum allowable circuit pressure
- 10-100 cm H_2O

Sigh
- Once every 100 breaths or seven minutes
- 1.5 times the VT or the I time
- High Pressure limit increases 1.5 times

Manual Breath
- Manual trigger for breath at set parameters

Alarms

High Peak Pressure
- Audible/visual alarm, flow stops, exhalation valve opens
- If pressure resets < 26 cm H_2O, normal ventilation resumes
- If pressure remains > 26 cm, ventilation remains suspended and audible/visual alarms sound
- Patient can breathe through anti-suffocation valve
- Normally set 15-20 cm above baseline Peak Pressure
- Range 1-100 cm H_2O
- Silence interval: 30 seconds

Low Peak Pressure
- Active in Control and Assist Control only
- Audible/visual when airway pressure fails to exceed the alarm setting during inspiration
- Range 2-50 cm H_2O
- Silence interval: 30 seconds
- Alarm Silence / Reset
- Temporarily disable some audible alarms
- Resets audible/visual indicators when condition ceases
- Can't pre-silence and alarm

I:E Ratio
- Audible/visual when I time > 50% of the total breath time as defined by the Breath Rate control; cannot be silenced
- Apnea Alarm
- Audible/visual when the period between two consecutive inspirations > 20 seconds
- Initiates apnea backup ventilation

Disconnect
- Audible/visual if a pressure increase of 2 cm H_2O or greater is not detected during inspiration
- Silence interval: 30 seconds

Ventilator Inoperative
- Audible/visual when normal operation ceases
- Breathe room air if spontaneous breathing is present recoverable

Loss of external power or voltage out of range
- Mode switch temporarily set to Off
- non-recoverable
- Software or CPU problem
- External Power Low/Fail
- Audible/visual which can be silenced
- Switches to internal battery

Battery Low/Fail
- Silence interval: 5 minutes
- PEEP Not Set
- Monitored PEEP value deviates more than 5 cm from manually set value
- Silence interval: 30 seconds

Transducer Calibration
- Self test shows baseline pressure +/-2 cm H2O from zero
- Calibrate ventilator

Monitors
- Airway Pressure
- Real time bar graph
- Range: 10-100 cm H2O
- Monitor Display
- Breath rate
- Flow
- High pressure alarm
- Low pressure alarm
- Manual PEEP
- I time
- VT

Peak Inspiratory Pressure (PIP)
- Peak pressure for last cycle (except spontaneous)
- Range 0-100 cm H2O

Mean Airway Pressure (MAP)
- Mean pressure over last cycle
- Range 0-100 cm H2O

Airway Pressure (Paw)
- Current airway pressure

Ventilation Limits
- Breath Rate/I Time
- Breath rate and I time must allow for a minimum 300 millisecond exhalation time

VT /Flow/Breath Rate
- VT, flow, and breath rate must allow for a minimum
- 300 millisecond exhalation time during volume ventilation

Ventilator Procedures
1. Ensure ETT placement
2. Provide initial support
3. Manual ventilation
4. PEEP
5. Cardiovascular stabilization
6. Stress of intubation
7. Delay in successful intubation
8. Drugs
9. Ensure appropriate ventilator settings
10. Establish baselines

Ventilator Procedures
- Set up ventilator
- FiO2
- Select mode
- Set/actual respiratory rate
- Set/actual tidal volume
- Set flow rate
- Connect ventilator
- Peak inspiratory pressure
- Set PEEP
- Set pressure alarms
- Patency of circuitry and all connections
- Monitor

Ventilator Procedures
- In case of instability or mechanical difficulty, disconnect the ventilator and use manual ventilation.

Vt Changes
- Decreased VT
- Malfunction of transducer
- Leak from chest tube
- Leak in circuitry, tube, cuff, etc.
- Increased VT
- Altered settings
- Transducer malfunction

Peak Airway Pressures

Increased pressure

Sudden
- Pneumothorax (tension)
- Plugs

- Intubation/migration of ETT to right main stem
- Kinked ET/ventilator tube
- Cuff herniation
- Position change

Gradual or intermittent
- Coughing or bucking
- Increased secretions
- Pulmonary edema
- Bronchospasm
- Water in tubing
- Atelectasis
- Pneumonia
- Abdominal distension

Decreased pressure
- ET tube displaced
- Cuff deflation / leak
- Leak in circuit or airway pressure tubing
- Tube too small (uncuffed tube)
- Chest tube
- Tracheal-esophageal fistula
- Portable Ventilators

Complications
- Airway trauma
- Barotrauma
- Machine failure
- Hypotension
- Pneumothorax
- Tension pneumothorax
- Atelectasis
- Pulmonary infection
- GI malfunction
- Renal malfunction
- CNS malfunction
- Psychological trauma

Portable Ventilators
- Transport considerations
- Compatibility with unit and power supply
- Power failure
- Occlusion/plug
- Bucking
- Barotrauma
- Patient need for sedatives/relaxants

(Fill in the blank with an up or down arrows)

Respiratory Acidosis

pH _____   PaCO2 _____

Respiratory Alkalosis

pH _____   PaCO2 _____

Metabolic Acidosis

pH _____   PaCO2 _____

Metabolic Alkalosis

pH _____   PaCO2 _____

Respiratory Parameters

The respiratory system's primary function is:

- Transport of oxygen to and removal of carbon dioxide from the cells
- This is achieved by the following processes:
    - Pulmonary ventilation
    - Movement of air between the atmosphere and the alveoli
    - Diffusion of the movement of oxygen and carbon dioxide between the alveoli and blood
    - Transport of oxygen to the peripheral tissues
    - Exchange of carbon dioxide at the cellular level
    - Return of carbon dioxide to the lungs
    - Regulation of ventilation

Explain the movements of the diaphragm.

- During inspiration, the diaphragm contracts and flattens increasing the size of the thorax
- During expiration, the diaphragm relaxes and the lungs recoil to decrease the size of the chest
- The diaphragm rises to the 4th intercostal space during expiration and extents to the 10th or 12th intercostal space during inspiration

What keeps the lungs fully expanded?

- The pressure within the plural cavity must always remain slightly negative, < 4 mmHg in relation to the atmospheric pressure.
- The lungs have an elastic tendency to recoil while the thorax cavity has an elastic tendency to expand.

How much fluid is contained between the two layers of the pericardial sack?

- 3 – 5 ml of lubricating fluid

Minute Volume
- RR x TV= MV
- Normal is 6-10L/min

Tidal Volume
- 6-10 ml/kg.

Rate
- 8-20/min

Flow rate
- 40-100 L/min

Inspiratory to Expiratory Ratio
- 1:2 or 1:3

Oxygen Concentration
- 40-100%

Normal Electrolyte Parameters	
Sodium Na+	135-145mEq/L
Potassium K+	3.5 - 5.5 mEq/L
Calcium Ca++	8.8 - 10.4 mg/dL
Magnesium Mg+	1.5 - 2.5 mEq/L
Chloride Cl-	95 - 105 mEq/L
Phosphorus	3.0 - 4.5 mg/dL
Bicarb, HCO3-	21 - 28 mEq/L
BUN	6 - 23 mg/dL
Creatinine	0.6 - 1.4 mg/dL
Glucose	70 - 110 mg/dL
Serum Osmo	285-295 mOsm/kg
Carbon Dioxide	24-30 mEq/L

Review Questions

1. Complications of intubation include all of the following, except:
    a. gastric distension
    b. arytenoids dislocation or avulsion
    c. bronchospasm
    d. pyriform sinus perforation

When preparing a patient in a rural ICU for transport, you look at the CXR. You note a ground glass appearance on the chest film. The patient is on a vent with settings at FiO_2 .80, TV900 cc, rate of 16 with 5 of PEEP. ABG results show a pH of 7.34, PO_2 of 76%, CO_2 of 38, and HCO_3^- of 24.

Daddy and his "Little Flight Nurse", age 1.

2. What pulmonary condition do you suspect?
    a. pneumothorax
    b. pulmonary edema
    c. ARDS
    d. cor pulmonale

3. Management of this patient would include:
    a. increasing rate to 20
    b. increasing the PEEP
    c. rapid needle decompression
    d. administering Lasix

4. A shift to the right on the oxyhemoglobin dissociation curve is caused by:
    a. alkalosis
    b. hyperthermia
    c. hypothermia
    d. decreased CO_2 levels

5. Indications for intubation of the asthmatic patient include:
    a. vital capacity above the level of tidal volume
    b. pH above 7.2
    c. pCO_2 > 55 mmHg.
    d. pO_2 < 80mm Hg.

Aircraft Safety, Survival, CAMTS

- Air Medical Safety issues
- Performing a Safety checklist
- Safety procedures during flight
- CAMTS / PAIP
- Survival skills
- Safety

What we do affects our crew, ancillary staff, public service members, bystanders, and most importantly...the patient.

We need to develop "safety consciousness" Developed through training, repetition, and complete familiarity with equipment and procedures.

Regulated and mandated by OSHA, and FAR's
- Safety knowledge
- Fire extinguisher usage
- Location of survival kit
- Flotation devices
- Emergent egress procedures
- ELT location and activation
- Fuel shutoff
- Battery disconnect
- Fuel spillage
- FAR Part 41, 61, 91, & 135
- General Operating and flight rules

For operating aircraft within U.S. airspace

FAR Part 135
Specifies rules for air taxi and commercial operators

Most Air Medical programs are Part 135 ops

Safety components required
- Safety Officer
- Communication methods (website, email, bulletin board, etc)
- A process of reporting to identify issues and a action plan
- Annual Safety Training
- Safety evaluations
- Hearing protection (decibel loss at >90)
- Clothing (Flight suit, Nomex, undergarments, reflective stripes, etc)
- Scene safety training

Accident rates
- HEMS accident rates far outweighs general aviation helicopter operations
- Increases each year since 1972
- From 1980-1985, estimated between 12.34 / 100,00 flights
- General aviation / air-taxi was 6.2/100,000
- In 1982, highest of all...16 in one year
- The NTSB concluded that of 59 HEMS accidents in 1988, weather was the most common and most concerning factor
- In 1999-2001, there were 121 accidents, again weather! This is a 10% increase since the 80's

CAMTS Requires...
- Safety Committee
- Composed of various members of the program
- Addresses hazard-reporting, then channeled.
- Also looks at CQI / QM, and performance improvement
- Action cycle should be implemented

Also discussed; weather min's, communications, training, etc

Along came CAMTS
- Developed in 1990
- Has recommended weather minimums
- Weather is the #1 cause of accident rates
- Any crew member can refuse a flight (3 to go, 1 to stay)
- They also address other safety issues and "best practices"

CAMTS defines...
- A Critical Care mission must have either an RN or specially trained Physician. (Can be RN-RN, RN-EMT-P, RN-RT, RN-MD, etc)
- Specialty mission must have at one AMC

- ALS mission is EMT-P/RN or EMT-P/EMT-P
- BLS (As stated)

ASTNA on Safety
- Formerly the NFNA
- Established standards for Flight Nursing
- They addressed crew rest and scheduling issues
- The right to refuse a flight if you are uncomfortable
- Addresses physical and chemical restraints for combative patients
- Hot-loading and un-loading after proper training
- Aircraft safety, i.e.: energy attenuating seats, addition of shoulder belts to lap belts, and crash resistant fuel systems

ASTNA position paper...
- Securing equipment during flight
- Use of seat belts and shoulder harnesses
- Proper patient securement
- Use of night lighting
- Isolation of pilot and controls from pt.
- Active participation of safety briefings with pilot

Emergent procedures

ELT activated at _____

Emergency freq _____

Post mission briefings

Position reporting

Communication terminology

Protective gear (Helmets, flame resistant flight suit, reflective stripes, protective footwear)

PAIP

Should be implemented after _____ minutes after failure to give position report

"Observe sterile cockpit during critical phases of flight"
Sterile cockpit: Silence during _____,

_____, and _____.

Eyes should be _____ the aircraft during taxi and landing

If a FCM identifies a hazard, immediately

notify _____.

During FW ops, must be belted during

_____, taxi, and landing.

Pilot Qualifications

- Minimum of _____ flight ready pilot's permanently assigned per 24 hr. single ship operation.

- Commercial Instrument RW rating of _____ hrs, with 1000 hrs as PIC & 100 hrs at night as PIC.

- Initial training includes terrain & weather, orientation to hospital, infection control, med systems on a/c, pt. loading and unloading, EMS / Public service agencies,

Instrument meteorological conditions (IMC)
- Recovery procedures by ref to Instruments or IFR currency
- 5 hrs area orientation w/ 2 hrs at night as PIC prior to HEMS missions
- Recurrent 135 training
- IMC recovery procedures
- Instruments / IFR
- Operating procedures
- Roles and Responsibilities
- Area / Weather / Terrain

Weather Minimums (Will probably change when new 7th edit CAMTS revisions go into effect)

Local

Day

Night

Cross County (Defined as 50 nm from base)

Day (notice Local night)

Night

Terms to remember...

VMC

I-IMC

VFR

IFR

During an Emergency...

-

-

-

-

Post Crash Procedures

ELT activation around

_____?

Emergency transmission frequency?

If pilot is incapacitated, disengage throttle, fuel, and then battery.

Do not exit until all motion stops.

Exit and meet at _____.
Secure

A.

B.

C.

D.

Rule of 2's

2 hrs

2 days

2 weeks

- CRM Training Facilitates open communication between crew

Addresses Preliminary events
- Pre-flight events
- Flight related events
- Emergency related events
- Survival related events
- In-flight events
- LZ issues
- A/C issues
- Teamwork
- Communication skills
- Decision making
- Workload mgt.
- Situational awareness
- Preparation and planning
- Cockpit distractions
- Stress mgt.
- Speaking up when there is a potential problem

CRM Continued

1. Address the person by name
2. State own emotion
3. State perceived or real problem
4. Offer a solution
5. Obtain recommendation and agreement

Review time...

1. RW weather minimums clearly state...

2. We meet where after a crash?

3. PAIP activated within ___ min after last position report?

4. Space between flight suit and undergarments?

5. Post crash again...first priority is ___?

6. Required by CAMTS? (Helmets? Nomex? Their own dispatch Ctr?)

7. Eyes where during takeoff and landings?

8. Voluntary guidelines?

   CAMTS?
   AAMS Weather mins?
   FAA?
   OSHA?
   NTSB?

9. Who is responsible for safety?

10. Sterile cockpit is what?

11. #1 killer of people in survival?

12. Emergent landing in water...when do you inflate flotation devices?

13. In water, when do you exit the a/c?

14. Most common cause of death from injury following an accident?

15. Emergency freq?

16. Temporary shelter best location? (stream, trees, near a/c, dead trees, edge of clearing, under cliff, etc)

17. The most important characteristic of survival is what?

Review Questions

1. Transport program safety is the responsibility of whom?
   a. The program or administrative director
   b. The designated safety officer
   c. All members of the transport team
   d. The pilot or driver in command

2. Sterile cockpit is defined as:
   a. Ensuring that all patient care equipment has been disinfected appropriately
   b. A conversation of a social nature occurring during the aircraft approach
   c. Conversation with the communications center during the transport process
   d. Conversation- restricted to the comments needed for the safe operation of the aircraft on all take-offs and landing's.

3. The number one killer of people in survival situations is:
   a. Hunger
   b. Thirst
   c. Fatigue
   d. Cold

4. The definition of "local flying area" is determined by:
   a. The part 135 certificate holder
   b. The program management
   c. The base hospital referral center
   d. The local flight standards district office

5. Your flight suit should be fitted to provide ___ space of insulation.
   a. 1/8"
   b. 1/4"
   c. 1/2"
   d. 1"

6. You are transporting a camera man for a local news station. During an approach to landing, he shouts out, "Hey, get close to our building so I can get some cool close-up shots." Is this scenario ok? Why?

Stressors of Flight

Hypoxia represents oxygen deficiency in body tissues sufficient to cause impairment of physiological functions.

Hypoxic Hypoxia -

Hypemic Hypoxia -

Stagnant Hypoxia -

Histotoxic Hypoxia -

Hypoxia - four stages
Increased altitude = decreased O2 in blood

Indifferent
- Sea Level to 10,000 ft
- Increased HR, RR
- Night vision decreases 28% (Guess where you will see this again?)

Compensation
- 10,000 - 15,000 ft
- Increased HR,RR, BP
- Night vision decreases 50%
- Impairment of task performance occurs

Disturbance
- 15,000 - 20,000ft
- Dizziness, sleepiness, tunnel vision
- Cyanosis
- Loss of muscle coordination

Critical
- 20,000 - 30,000 ft
- Mental confusion and incapacitation
- Unconsciousness in a few minutes

Time of Useful Consciousness

Elapsed time from exposure to an oxygen deficient environment to the point where deliberate function is lost. i.e.: loss of cabin pressure

< 18,000 ft	30 minutes
25,000 ft	3-5 minutes
30,000 ft	***90 seconds (hint)***
35,000 ft	30-60 seconds
> 40,000 ft	< 15 seconds

Effective Performance Time

The amount of time a Flight Crew member is able to perform their useful flying duties in an inadequately oxygenated environment

Barometric Pressure

(As altitude increases atmospheric pressure decreases and gases expand)

Barotitis Media -

Barosinusitis -

Barodentalgia -

Respiratory System –

Medical equipment -

Thermal Changes

As altitude increases ambient temperature decreases *2 degrees C for every 1,000 feet*

ICE IS A KILLER OF ANY AIR MEDICAL PROGRAM. With any precipitation, clouds, or narrow dew point spread, ice can form on the wings, causing a disruption of airflow, thus drastically reducing lifting forces of the aircraft.

Hypothermia enhances hypoxia, as heat loss is usually through evaporation and conduction both hypothermia and hyperthermia. This can cause an increase in metabolic rate causing increased oxygen demands

Humidity/Dehydration
As altitude increases the amount of moisture in air droplets decreases causing some concern in high altitude long distance flights

Signs of dehydration

-

-

-

Noise

- The longer the exposure to noise and more intense the noise, the greater the damage
- Individual variation as to tolerance
- Can interfere with patient care, may impede communications between crew members

Flicker Effect

Vibration

Most common sources of vibration are the power plant and turbulent air at altitude. (Hearing loss >90 dbl)

- may increase metabolic rate
- may cause fatigue or motion sickness
- may worsen hypothermia
- may interfere with patient monitoring
- additional patient padding may be beneficial

Fatigue

Can be the end product of any or all the contributing factors of flight induced stress

Can be self imposed or enhanced by

D -

E -

A -

T -

H -

Gravitationnel Forces

- Usually not a significant factor, but are applied to the body on ascent and decent and during change in speed and direction enhanced by patient positioning
- With head toward rear of a/c, blood pools in upper body
- With head toward front of a/c, blood pools in lower body

Helpful Flight Crew Hints

- a well trained crew anticipates hazards
- hypoxia is the greatest potential hazard
- barometric pressure changes cause pain
- gas expansion should be anticipated
- minimize the effect of self-imposed stress

Helpful Patient Care Hints

- switch all IV containers
- discontinue all unnecessary IV lines
- insert prophylactic indwelling lines
- pre-medicate and sedate
- low hemoglobin = blood infusion
- get all medical records and X-rays
- empty all indwelling lines/tidy up

Emergency Medical Treatment and Active Labor Act (EMTALA) of COBRA - 1986

Emergency medical condition that poses a serious health threat or threatens impairment of organs or bodily function

Case Review 1

57 y.o. female in the ICU at the local community hospital being transferred to John Hopkins Medical Ctr. Pt with unstable AMI.

Past medical history:
1 previous MI, CHF
Patient vital signs:

BP 108/68; RR 16; HR 80 ; SaO2 97% on NRB

Pt on NTG gtt., Heparin, and Amiodarone / Milrinone infusions

Flight Crew Considerations:

- mode of transport/length of transport
- space/equipment considerations
- pharmacological support
- cardiopulmonary response to hypoxia
- fluctuations in encapsulated air pressures
- protect patient from thermal variants
- monitor effects of noise and vibration

Case Review 2

- 35 y.o. male unrestrained driver MVA vs. tree.
- Lengthy extrication was just completed by BLS crew on your arrival.
- Patient on full backboard with C-collar in place.

Physical Exam

- head lacerations\contusions
- deformed left arm
- mid-sternal chest tenderness

Vital Signs
BP 98/62; RR 30; HR 124; SaO2 88% on NRB

Flight Crew Considerations:

- LZ considerations and scene safety
- assess patient's ABCs and C-spine control
- assess patient's LOC, secondary survey, v\s
- perform any on-scene interventions
- secure\package patient for flight
- monitor for altitude related metabolic changes
- notify receiving facility

Fly safe…….... and enjoy the view

Thoracic Trauma

Major signs and symptoms
- Immediate life-threatening injuries

Pathophysiology and management
- Open pneumothorax
- Tension pneumothorax
- Massive hemothorax
- Flail chest
- Cardiac Tamponade

Cardiac involvement with blunt injury

Other thoracic injuries

Thoracic injury is common.
- 50% of multiple trauma
- 25% of trauma deaths

Potentially fatal thoracic injuries saved by rapid recognition and intervention.
- Many require surgical intervention

Chest Anatomy
- Many vital organs are crowded into this area. Trauma to this area is often life-threatening.
- Bony cavity formed by 12 pairs of ribs, which join posterior with thoracic spine and anteriorly with sternum.
- Intercostal neurovascular bundle runs along inferior surface of each rib.
- Inner side of cavity and lung itself are lined with thin layer of tissue, pleura. The space between pleural layers is normally only a potential space.
- One lung occupies each thoracic cavity.
- Between two cavities is mediastinum, which contains heart, aorta, superior and inferior vena cava, trachea, major rhonchi, and esophagus. (High potential for life-threatening injury because of vital cardiovascular and tracheobronchial structures within this area.)
- Spinal cord is protected by vertebral column.
- Diaphragm separates thoracic organs from abdominal cavity. Upper abdominal organs, including spleen, liver, kidneys,

pancreas, and stomach, are protected by lower rib cage.

Mechanism of Injury

Blunt
- Direct compression
  - Fracture of solid organs
  - Blowout of hollow organs
- Deceleration forces
  - Tearing of organs and blood vessels

Penetrating
- Direct trauma to organ and vasculature
- Energy transmitted from mass and velocity

Tissue Hypoxia
- Inadequate oxygen delivery
- Hypovolemia
- Ventilation/perfusion mismatch
- Pleural pressure changes

Pump failure
- Inadequate oxygen delivery to tissues secondary to airway obstruction
- Hypovolemia from blood loss
- Ventilation/perfusion mismatch from lung parenchymal injury
- Changes in pleural pressures from tension pneumothorax
- Pump failure from severe myocardial injury

Thoracic Trauma
- Shortness of breath
- Chest pain
- Hemoptysis
- Cyanosis
- Neck veins distended
- Tracheal deviation
- Asymmetrical movement

Signs and symptoms
- Chest wall contusion
- Open wounds
- Subcutaneous emphysema
- Shock
- Tenderness, instability, crepitation (TIC)

Breath sounds abnormal
- Major symptoms are shortness of breath and chest pain.
- Mechanism of injury is also a sign of chest injury.

Trauma Primary Survey
"Deadly Dozen"
1. Airway obstruction
2. Open pneumothorax
3. Flail chest
4. Tension pneumothorax
5. Massive hemothorax
6. Cardiac Tamponade

Trauma Secondary Survey
"Deadly Dozen"
7. Myocardial contusion
8. Traumatic aortic rupture
9. Tracheal or bronchial tree injury
10. Diaphragmatic tears
11. Esophageal injury
12. Pulmonary contusion

Airway obstruction
- Secondary hypoxia
  - Common cause of preventable death
  - Foreign body, tongue, aspiration
- Always assume cervical spine injury

NOTE: Management of airway has been discussed in airway, so nothing further will be added here other than to stress its importance.

Open pneumothorax
- "Sucking chest wound"
  - Air enters pleural space
  - Ventilation impaired
  - Hypoxia results
- Signs and symptoms
  - Proportional to size of defect

Primary "Deadly Dozen"
- Normal ventilation involves negative pressure being generated inside chest by diaphragmatic contraction.
- As air is drawn through upper airway, lungs expand.
- With a large open chest wound (larger than trachea or about size of patient's little finger), path of least resistance for airflow is through chest wall defect.

- Air going in and out of this opening makes a sucking sound, and bubbles on expiration.
- This air will enter only pleural dead space. It will not enter lung and therefore will not contribute to oxygenation of blood. Ventilation is impaired, and hypoxia results.

Open pneumothorax
- Close chest wall defect
- Load-and-go
- Use impervious material taped on three sides.
- The Asherman chest seal can be used to seal a sucking chest wound or can be placed over a decompressing needle.

Flail chest:
- Three or more adjacent ribs are fractured in at least two places, resulting in segment of chest wall that is not in continuity with thorax.
- A lateral flail chest or anterior flail chest (sternal separation) may result.
- With posterior rib fractures, heavy musculature usually prevents occurrence of a flail segment.
- If patient is breathing spontaneously, flail segment moves with paradoxical motion relative to rest of chest wall.
- Multiple rib fractures with or without flail chest can cause hypoxia from mechanical ventilatory problems as well as pulmonary contusion. Patient, especially if older, must be closely monitored for hypoxia and respiratory failure.
- Monitoring with pulse oximetry and capnography is very helpful.
- Assist ventilation
- Possible intubation
- Load-and-go
- Stabilize flail segment
- Monitor for:
  - Pulmonary contusion
  - Hemothorax
  - Pneumothorax
- Intubation and positive pressure ventilation are best stabilization. This is usually not possible, as patient is usually awake with an intact gag reflex.

- Flail may contribute to development of pulmonary contusion, hemothorax, pneumothorax.
- Consider intubation early in order to provide positive end-expiratory pressure.
- Continuous positive airway pressure (CPAP) could be used in non-intubated patient.

Tension pneumothorax

- Dyspnea
- Anxiety
- Tachypnea
- Distended neck veins
- Tracheal deviation (rare)
- Breath sounds diminished
- Hypertympany if percussed
- Shock with hypotension
- Tension pneumothorax is a circulatory (obstructive) emergency.
  - Occurs when a one-way valve is created from either blunt or penetrating trauma. Air can enter but not leave pleural space.
- This causes an increase in intrathoracic pressure, which will collapse affected lung and will then exert pressure on mediastinum.
- This pressure will eventually collapse superior and inferior vena cava, resulting in a loss of venous return to heart.
- A shift of trachea and mediastinum away from side of tension pneumothorax will also compromise ventilation of other lung, although this is a late phenomenon and usually cannot be detected except by x-ray.

Tension pneumothorax

- Decompress affected side
  - Respiratory distress and cyanosis
  - Loss of radial pulse
  - Decreasing level of consciousness
- Load- and- go
- Loss of breath sounds on one side does not make a diagnosis of tension pneumothorax.
- A needle decompression is a temporary, but life-saving, measure.

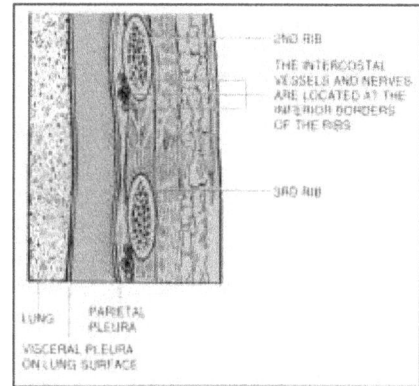

Massive hemothorax

- Blood in pleural space is a hemothorax.
- A massive hemothorax occurs as a result of at least a 1,500 cc blood loss into thoracic cavity. Each thoracic cavity may contain up to 3,000 cc of blood.
- As blood accumulates within pleural space, lung on affected side is compressed. If enough blood accumulates (rare), mediastinum will be shifted away from hemothorax. The inferior and superior vena cava and contralateral lung are compressed. Thus, ongoing blood loss is complicated by hypoxemia.

Signs and Symptoms
- Anxiety and confusion
- Neck veins
- Flat:
  - hypovolemia
- Distended:
  - mediastinal compression
- Breath sounds decreased
  - Hypotympany if percussed / Shock

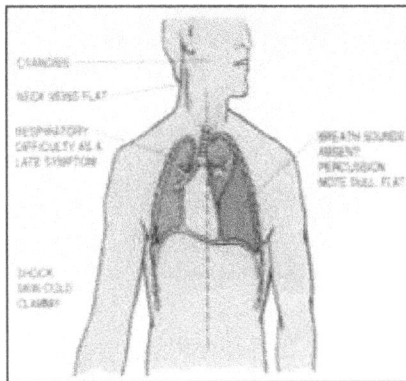

Massive hemothorax

- Test question…do not evacuate more than 500cc per affected side at a time
- Load-and-go
- Treat for shock
- Fluid administration
  - Titrate to peripheral pulse (90–100 mmHg)
- Monitor for:
  - Hemopneumothorax
- Fluid administration may increase bleed.

Cardiac Tamponade

- Beck's triad
  - **Hypotension**
  - **Neck veins distended**
  - **Heart sounds muffled**
- Paradoxical pulse
- Breath sounds equal
- The pericardial sac is an inelastic membrane that surrounds heart. If blood collects rapidly between heart and pericardium from a cardiac injury, ventricles of heart will be compressed.
- A small amount of pericardial blood may compromise cardiac filling. As compression of ventricles increases, heart is less able to refill, and cardiac output falls.
- Load-and-go
- Treat for shock
- **Fluid administration**
  - **Titrate to peripheral pulse (90–100 mmHg or MAP of >70)**
- Monitor and treat dysrhythmias
- Monitor for:
  - Hemothorax
  - Pneumothorax

- Fluid administration may increase bleeding.

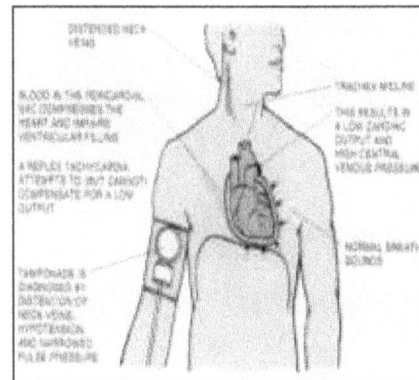

Myocardial contusion
- Most common cardiac injury
  - Blunt anterior chest injury
- Same as myocardial infarction
  - Chest pain
  - Dysrhythmias
  - Cardiogenic shock (rare)
- Treat as cardiac Tamponade
- Myocardial contusion is a potentially lethal lesion resulting from blunt chest injury.
- The chest pain may be difficult to differentiate from associated musculoskeletal discomfort that patient also suffers as a result of injury.

Traumatic aortic rupture
- Most common cause of immediate death
  - motor-vehicle collisions or falls from heights
  - 90% die immediately
- Scene Size-up and history extremely important
  - No obvious sign of chest trauma
  - Hypertension in upper extremities and hypotension in lower extremities (rare)
- Traumatic thoracic aortic tears usually are due to deceleration injury with heart and aortic arch moving suddenly anteriorly (third collision), transecting aorta where it is fixed at ligamentum arteriosum.
- In 10% of patients who do not exsanguinate promptly, aortic tear will be contained temporarily by surrounding tissues and adventitia. However, this will usually rupture within hours unless surgically repaired.

NOTE: Hypertension in upper extremities and hypotension in lower (as assessed by pulse strength) are rare.

Tracheal or bronchial tree injury
- Subcutaneous emphysema
  - Chest, face, neck
- Ensure adequate airway
  - Cuffed ET tube past site of injury
- Monitor for:
  - Pneumothorax
  - Hemothorax
- NOTE: This may be one occasion when a main stem intubation would be beneficial.

Diaphragmatic tear
- Severe blow to abdomen
- Herniation of abdominal organs
  - More common on left
  - Breath sounds diminished
  - **Bowel sounds auscultated in chest** (rare)
  - **Abdomen appears scaphoid (test question for sure)**
  - **Scaphoid: "sucked in," comma shaped.**
- Tears in diaphragm may result from a severe blow to abdomen.
- A sudden increase in intra-abdominal pressure, such as a seat-belt injury or kick to abdomen, may tear diaphragm and allow herniation of abdominal organs into thoracic cavity.
  - Occurs more commonly on left than right, since liver protects right hemidiaphragm.
- Blunt trauma produces large radial tears in diaphragm.
- Penetrating trauma may also produce holes in diaphragm, but these tend to be small.

Esophageal injury
- Penetrating trauma
- Difficult to assess in field
- If unrecognized, may be lethal

Pulmonary contusion
- Common from blunt trauma
- Hours to develop
- Marked hypoxemia

Impaled object
- Do not remove
- Gently stabilize object
- Avoid movement
- A very common chest injury resulting from blunt trauma, a pulmonary contusion takes hours to develop and rarely develops during prehospital care, unless very long transport or delayed discovery of victim occurs.
- Contusion of lung may produce marked hypoxemia.

Management consists of
- Intubation and/or assisted ventilation if indicated
- Oxygen administration
- transport
- IV insertion

Traumatic asphyxia
- Severe compression
  - Ruptures capillaries
  - Cyanosis above crush
  - Swelling of head, neck
  - Swollen tongue, lips
  - Conjunctival hemorrhage

Simple pneumothorax
- Fractured ribs
- Pleuritic chest pain
- Dyspnea
- Decreased breath sounds
- Hypertympany if percussed
- Monitor for:
  - Tension pneumothorax

Sternal fracture
- Significant blunt trauma to anterior chest
- Signs of fracture on palpation
- Myocardial contusion presumed

Simple rib fracture
- Most frequent chest injury
- Monitor for:
  - Pneumothorax
  - Hemothorax
- Pneumothorax caused by accumulation of air within potential space between visceral and parietal pleura.
  - Lung may be totally or partially collapsed as air continues to

accrue in thoracic cavity. In a healthy patient, this should not acutely compromise ventilation, if a tension pneumothorax does not evolve.
- Patients with less respiratory reserve may not tolerate even a simple pneumothorax.

Chest injuries common

Often life-threatening
- Require prompt recognition
- Require prompt intervention
- Frequently require load-and-go

Airway and oxygenation always priority
Frequent Ongoing Exams

Thoracic Review Questions

1. Your patient was trapped under a tractor for almost 8 hours. Once extricated, he is most likely to experience;
   a. Tension Pneumothorax
   b. Massive hemothorax
   c. Aortic rupture
   d. Rhabdomylosis

2. How much blood should be evacuated from a hemothorax?
   a. As much as needed to stabilize the patient
   b. No more than 500 cc's per side
   c. None, if the patient is stable, as this problem will resolve itself
   d. No more than 250 cc's, as this could cause the patient to go into shock

3. Chest tube insertion location is done at the following:
   a. between the 2nd and 3rd intercostal space, mid-clavicular, on the superior aspect.
   b. At the margin of the 5th intercostal space, anterior-axillary.
   c. At the point just distal to the lower lateral ribs.
   d. Between the 5th and 6th intercostal space, anterior-axillary, inserting a gloved finger into prior tube insertion.

4. Your patient was struck from behind while driving. He should be evaluated for;
   a. Hangman's fx
   b. Coup-contrecoup injury patterns
   c. Frontal impact injuries
   d. All of the above

5. A patient presents with a scaphoid abdomen, diminished bowel sounds, and unilateral dullness to percussion. What is the likely cause of his injuries?
   a. Tracheobronchial rupture
   b. lacerated transverse bowel
   c. Diaphragmatic rupture
   d. Constipation and bowel obstruction

Abdominal Trauma

Basic abdominal anatomy
- How abdominal and chest injuries are related

Blunt and penetrating injuries
- Complications associated with each
- Treatment for protruding viscera
- Relationship of exterior and underlying injuries

Possible intra-abdominal injuries
- History, physical examination, mechanism of injury

Abdominal ALS interventions

Difficult to evaluate
- Attention to scene and mechanism of injury

Major cause of preventable death
- Hemorrhage
- Anticipate shock: immediate or delayed
- Require surgical intervention
- Infection
- Gross contamination prevention
- Important information from scene: Note circumstances surrounding injury.
  - Accurate but rapid assessment of scene will usually tip you off to possibility of abdominal trauma.
- Major cause of preventable death MUST be recognized and treated immediately.
  - Early recognition and treatment can prevent these deaths.
- Hemorrhage has immediate consequences; look for early shock in all abdominal-injury patients.
- Infection, which presents late, may be just as deadly.

Abdominal Regions

Thoracic abdomen

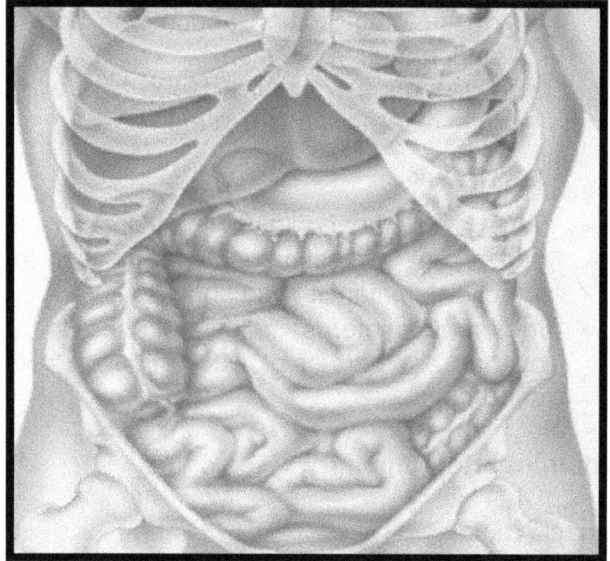

- Thoracic abdomen located underneath diaphragm and enclosed by lower ribs, which offer protection.
  - Contains liver, gallbladder, spleen, stomach, and transverse colon.
  - Any penetrating injury below 4th intercostals space may have penetrated abdomen.

Retroperitoneal

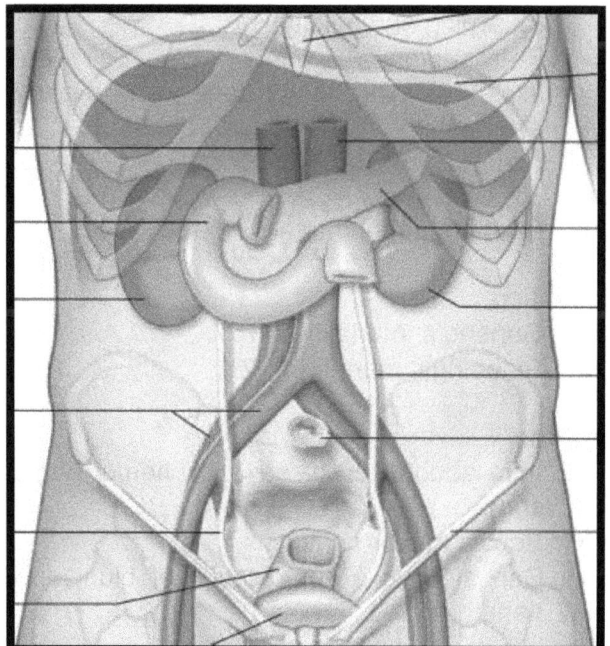

- Retroperitoneal abdomen is located behind thoracic and true abdomen.
  o Contains kidneys, ureters, pancreas, posterior duodenum, ascending and descending colon, abdominal aorta, and inferior vena cava.
  o Can conceal massive blood loss with little external signs (except shock).

True abdomen
- True abdomen contains small intestines and bladder. Intestinal injury can result in infection, peritonitis, and shock.
- In females, uterus, fallopian tubes, and ovaries are part of pelvic portion of true abdomen.

Abdominal Region Injury

Thoracic region
- Life-threatening hemorrhage: liver, spleen

True abdomen
- Infection, peritonitis, shock: intestines
- Severe hemorrhage with signs

Retroperitoneal abdomen
- Severe hemorrhage hidden: major vessels
- Injury in retroperitoneal produces different symptoms than in true abdomen.
- True abdomen can present with distension, tenderness, and tenseness (guarding and rigidity).
- Retroperitoneal injury can conceal exsanguinating hemorrhage with no early symptoms.

Blunt
- Most common: mortality 10–30%

Penetrating
- Gunshots: mortality 5–15%
- Stabbings: mortality 1–2%

Concern:
- Intra-abdominal bleed with hemorrhagic shock
- Sepsis and/or peritonitis
- Can also be a combination of blunt and penetrating.

- Gunshot wounds have higher mortality (up to 15%), due to higher rates of damage to abdominal viscera.
- Stabbings: approximately one-third require surgery.
- Causes of mortality: hypovolemic shock, injury to abdominal viscera.
- Sepsis and/or peritonitis are late causes of death.

Scene Size-up

Remember to lift and look!

Scene Size-up and mechanism can provide valuable clues to possible abdominal injury.

Blunt Abdominal

Mechanism
- Direct compression of abdomen
- Fracture of solid organs (spleen/liver)
- Blowout of hollow organs (intestines)
- Deceleration forces
- Tearing of organs and blood vessels

Accompanying injuries
- Head, chest, extremity: 70% MVC victims
- Direct compression of abdomen: fracture of solid organs (spleen/liver) and blowout of hollow organs (intestines).

Liver and spleen injury most common

Evidence of injury
- Often no or minimal external evidence
- Significant blood volume concealed in regions
- Seat-belt sign: 25% intra-abdominal

Pain or tenderness
- Often no pain or overshadowed by other pain
- Multiple lower rib fractures—patients notorious for having severe intra-abdominal injuries without significant abdominal pain.
- Seat-belt sign is a large abrasion over abdomen and/or upper neck.
- It is indicative of intra-abdominal injury in approximately 25% of cases.

Penetrating Abdominal

Mechanism

- Direct trauma to organ and vasculature
- Projectile and fragments
- Energy transmitted from mass and velocity

Caution:
- Vigorous fluid resuscitation may do more harm
- Penetrating injury often involves uncontrolled hemorrhage.
- Vigorous fluid administration may only worsen rate of hemorrhage.
- Projectile pathway not always obvious

Abdominal injury is chest; chest is abdominal
- Gluteal area in 50% of significant injuries
- Path of penetrating object might not be readily apparent from wound location.
- Any penetrating wound of chest may penetrate abdomen, and vice versa.
- Bullet may pass through numerous structures in different body locations.
- As discussed in Scene Size-up, ballistics information (caliber, velocity, trajectory, range, etc.) contributes to extent of injury and is helpful information.
- Gluteal area (iliac crests to gluteal folds, including rectum) is associated with up to a 50% incidence of significant intra-abdominal injuries.

Abdominal Assessment

Primary Survey: Abdomen
- Deformities
- Contusions
- Abrasions
- Punctures
- Evisceration
- Distension
- Tenderness
- Tenseness
- Keep high suspicion—rapid visual evaluation and palpation.
- Auscultation or percussion in field loses critical time, and little useful information is gained.

- If clothing removed to visualize injury, try to preserve important potential legal evidence by cutting around (rather than through) areas that have signs of possible penetration.

Signs and Symptoms

Splenic injury
- Referred left posterior shoulder pain **(DING DING DING!)**

Liver injury
- Referred right posterior shoulder pain **(GUESS WHERE THIS WILL BE?)**

Severe hemorrhage
- Distention, tenderness, tenseness
- Pelvic tenderness or bony crepitation
- Mechanisms for referred pain are not completely understood, but several methods/causes of referred pain have been identified. Visceral (organ) and somatic (skin, muscle, connective tissue) sensory nerves transmit pain signals to a spinal nerve ganglion where it is then transmitted to spinal cord and then to brain. Visceral pain is usually dull or poorly localized, and somatic pain is usually sharp and well-localized.
- Pain stimulation from a sensory nerve (visceral or somatic) that is compressed or damaged at or near its origin can be perceived as originating in additional areas innervated by injured sensory nerve.
- Pain stimulation from a damaged intervertebral disc can cause compression on nerve root coming from spinal cord at that level and be felt in additional regions served by compressed nerve.
- In addition, visceral pain stimulation can be felt in areas normally innervated by somatic sensory nerves. Visceral and somatic sensory signals converge in spinal cord.
- Signals from this level of spinal cord can be perceived as originating from somatic nerve: for example, irritation of diaphragm is signaled by phrenic nerve and can be perceived as pain in area above clavicle (Kerr's sign).

Stabilization
- Signs usually do not appear early.
- If present, injury is significant.

Assess and treat for shock.
- Lack of tenderness does not rule out injury with altered mental status and/or spinal injury at or above level of abdomen.
- Blunt trauma to abdomen with abdominal pain and/or tenderness probably has serious abdominal trauma and is likely to develop shock quickly (even if vital signs are initially normal).
- Load and go, preparing to treat development of hemorrhagic shock en route to hospital.
- Aggressive fluid resuscitation might dislodge protective clots and dilute clotting factors, which would lead to worsening hemorrhage. Administer fluid cautiously and check with medical direction.

Special Situations

Evisceration
- Do not push viscera back into abdomen.
- Gently cover with moistened gauze.
- Apply nonadherent material to prevent drying.
- If intestines are allowed to dry, they may become irreversibly damaged.
- Flex legs slightly at knees to take pressure off abdominal musculature.

Impaled object
- Do not remove.
- Uncontrollable hemorrhage
- Gently stabilize object.
- Avoid movement.

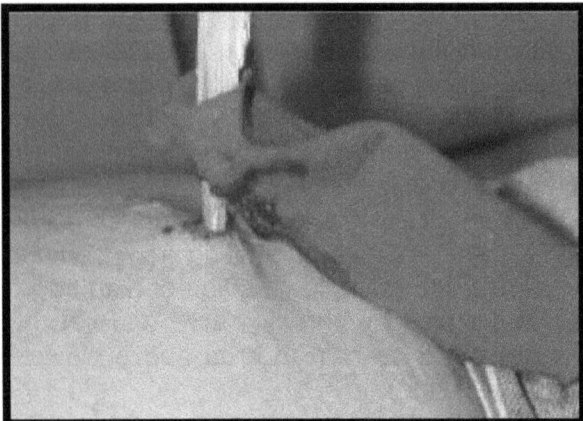

A piece of wood shot out of a planer at a furniture plant entered this worker's abdomen. He complains of pain, nausea. He is tachycardic at 120 bpm, and he has peripheral pulses.

Removal or manipulation may precipitate uncontrollable hemorrhage.
Never flex legs with impaled objects—causes additional soft-tissue injuries.

Abdominal Review Questions

What is the 3rd most common cause of traumatic death?

- Abdominal injuries, preceded by head and chest injuries
- Abdominal trauma results in a mortality rate of 13 to 15%
- Approximately 1/5th of all traumatized patients require abdominal surgery

What is the most common mechanism of blunt abdominal injury?

- Motor vehicle crash
- Firearms, stabbings, and physical assaults are more associated with penetrating abdominal trauma

Lap belt use has been associated with injury to what body structures?

- Hollow organs, particularly the small bowel and colon
- Lumbar spine
- Abdominal wall

Frontal impact crashes with a bent steering wheel and broken windshield are associated with what injuries?

- Spleen and liver injuries
- Head
- Chest trauma

What organs are most commonly injured by blunt trauma?

- Liver and spleen

What abdominal organs are most commonly injured by penetrating trauma?

- Liver, small bowl and stomach

Lower rib fractures are most commonly associated with injury to what organs?

- Spleen and liver

Pelvic fractures are most commonly associated with what intra-abdominal injury?

- Bladder laceration

T/F: If a patient has penetrating wounds at the nipple line they are considered to be at risk for intra-abdominal injury.

- True

Why is evaluation of blood loss in the abdomen difficult?

- Bleeding from structures in the retroperitoneum leads to hemorrhage in the retroperitoneum, which is very difficult to evaluate and diagnose

What are the classic signs and symptoms of intra-abdominal pathology?

- Pain
- Rigidity
- Guarding
- Spasm

What causes rebound tenderness and guarding in the abdominal muscles?

- Sudden movement of irritated peritoneal membranes against the abdominal wall

Why may injuries to the pancreas and surrounding tissues result in delayed signs and symptoms?

- Enzymes released into the retroperitoneum may cause chemical peritonitis, and may cause significant

tissue swelling which may not appear for several hours after the injury
- Patients with pancreatic or duodenal injury may complain of diffusive abdominal tenderness and pain radiating from the epigastric area to the back

What is *Kehr's sign*?

- Referred pain to the left shoulder following splenic rupture
- Blood collects under the diaphragm causing irritation of the phrenic nerve which innervates the diaphragm
- Pain is perceived along the course of the nerve, and is commonly located in the left subscapular region

Where may pain of a duodenal injury be referred?

- The patient may complain of testicular pain

How is a patient with a traumatic liver injury evaluated for non-operative management?

- Evaluate hemodynamic stability
- Absence of peritoneal signs
- Neurological integrity
- CT scan
- Degree of free intraperitoneal blood
- Absence of associated intra-abdominal injuries
- Need for 2 or more hepatic related blood transfusions
- CT scan documented improvement or stabilization with time

What are the signs and symptoms of hepatic injury?

- Upper right quadrant pain
- Abdominal wall muscle rigidity spasms or guarding
- Rebound tenderness
- Absent or hypoactive bowel sounds
- Signs or hemorrhage and/or hypovolemic shock

Fractures of the left 10th through 12th ribs are associated with injury to what organ?

- Spleen

- Splenic injuries range from laceration of the capsule to a non-expanding hematoma to a ruptured subcapsular hematoma or laceration

What is the most serious splenic injury?

- A fractured spleen or vascular tear, which results in splenic ischemia and massive blood loss

What are the signs and symptoms of splenic injury?

- Hemorrhage or hypovolemic shock
- Left shoulder pain (Kehr's sign)
- Upper left quadrant tenderness
- Abdominal wall muscle rigidity, spasm, or involuntary guarding

What hollow organ is most frequently injured?

- The small bowel
- Deceleration may lead to shearing, which results in avulsion or tearing

What are the signs or symptoms of hollow organ injuries?

- Peritoneal irritation, such as abdominal muscle wall rigidity, spasm, guarding, rebound tenderness, or pain
- Evisceration of the small bowl or stomach
- Diagnostic peritoneal lavage, which may show the presence of feces, bile, or food fibers

What is the most common injury to the kidney?

- Blunt contusion
- Rupture of the kidney is typically not associated with hypovolemia unless laceration of the renal artery has occurred

What are the sign and symptoms of renal injury?

- Ecchymosed over the flank
- Flank or abdominal tenderness solicited during palpation
- Gross or microscopic hematuria

T/F: The absence of hematuria does not rule out renal injury.

- True

What is the most common cause of injury to the bladder?

- Blunt trauma
- Most ruptures of the bladder occur as a result of pelvic fractures

T/F: Urethral trauma is more common in males than females.

- True
- Because the male urethra is longer and less protected
- The presence of an anterior pelvic fracture should raise the suspicion of a concomitant urethral injury
- Urethral injuries in females are usually associated with pelvic fractures

What is the most common mechanism of injury to the urethra in males?

- Skeletal trauma

Prostate (posterior urethral injury) is usually associated with what?

- Pelvic fractures
- Frequently associated with incontinence and impotence

What are the signs and symptoms of bladder and urethral injury?

- Suprapubic pain
- Urge but inability to urinate
- Hematuria, which may be microscopic
- Blood at the urethral meatus
- Blood in the scrotum
- Rebound tenderness
- Abdominal wall muscle rigidity, spasms or guarding
- Displacement of the prostate gland

What historical questions should be obtained in regards to a patient with potential abdominal trauma?

- Was the patient wearing restraints or protective devices?

- What is the location, intensity, and quality of the pain?
- Is nausea or vomiting present?
- Does the patient feel an urge to defecate or urinate?

What physical exam characteristics should be assessed when evaluating abdominal trauma?

- Inspect the abdomen for abnormal contour
- Inspect for abrasions, soft tissue injuries, and eccymosis over the flanks, which suggests splenic injury
- Inspect for gunshot or stab wounds
- Inspect the pelvic area for soft tissue bruising
- Inspect the peritoneum for hematomas, blood drainage from the urethral meatus, and/or vaginal or rectal bleeding
- Auscultate all four quadrants of the abdomen for bowel sounds
- Percuss the abdomen for hyper resonance
- Palpate the abdomen for guarding, rigidity, spasms, and localization of pain
- Palpate the pelvis for bony instability, asymmetry or pain
- Palpate the flanks for tenderness
- Palpate the anal sphincter for presence or absence of tone

What diagnostic procedure is best used for evaluation of retroperitoneal structures?

- CT Scan

What is the purpose of a flat plate lateral and upright abdominal radiographic studies in a trauma patient?

- Visualize foreign bodies and associated visceral damage
- Identify the path of penetrating objects
- Visualize free air in the abdomen

What studies are used to evaluate abdominal injuries?

- CT
- IVP
- Flat plate x-rays of the abdomen
- Cystogram/Urethrogram

- Diagnostic ultrasound or sonogram
- Angiography

In addition to the standard protocol of laboratory studies in a trauma patient, what specific laboratory studies should be used to evaluate abdominal injury?

- Serum amylase
- Liver function studies
- Analysis of urine, stool, or gastric contents for blood
- Pregnancy testing for females of childbearing age

Why has the diagnostic peritoneal lavage fallen out of favor in evaluating abdominal injuries?

- The diagnostic peritoneal lavage is not useful for identifying retroperitoneal bleeding and will often miss retroperitoneal injuries.

What precautions should be performed prior to performing a diagnostic peritoneal lavage?

- Decompress the bladder with an indwelling catheter and place a gastric tube to decompress the stomach

What is a diagnostic peritoneal lavage?

- Placement of a small catheter into the abdomen
- Aspiration of blood is considered a positive finding
- A liter of warmed ringers lactate solution or normal saline is rapidly infused into the catheter and then the lavage fluid is allowed to drain out of the abdomen via gravity and analyzed for the presence of red or white blood cells, bile, amylase, food fiber, or feces
- A DPL has a 98% accuracy rate in correctly identifying intra-abdominal bleeding

When does the American College of Surgeons Committee on Trauma recommend that a DPL be performed?

- Early to evaluate the severely injured hypotensive patient, especially if the abdominal examination is suggestive of an injury and the patient is unreliable or unresponsive

What are the contraindications to a peritoneal lavage?

- Absolute contraindication
  - When the decision has already been made to perform abdominal surgery
- Relative contraindications
  - When the patient has previous abdominal surgery increasing the potential for adhesions
  - When the patient has known cirrhosis of the liver
  - When the patient is extremely obese
  - When the patient has a known medical history of coagulopathy
  - Gravid uterus

What are the disadvantages of a diagnostic peritoneal lavage?

- Invasive
- Does not detect injuries to the diaphragm
- Does not detect injuries to the retroperitoneal structures

What are the disadvantages of ultrasound in the evaluation of abdominal injuries?

- Bowel gas and subcutaneous air distortion
- Does not detect diaphragm, bowel, and some pancreatic injuries

What are the disadvantages of CT scan in evaluation of abdominal injuries?

- Time delay
- May not detect diaphragm, bowel tract, and some pancreatic injuries

The patient is at risk for infection related to contamination of the peritoneal cavity by feces. What are the correct interventions?

- Maintain aseptic technique
- Cover open wounds with a sterile dressing
- Stabilize impaled objects
- Administer antibiotics as indicated
- Monitor temperature

- Check wounds for drainage
- Obtain blood cultures and laboratory studies
- Prepare for definitive care

The patient cannot eliminate due to urethral trauma. What are the correct interventions?

- Insert urinary catheter, unless contraindicated
- Monitor urinary output

Radiological Interpretation

Things we will see during Chest X-Ray Interpretation

1. Bony Structures
   (Shoulders, Humorous, & Chest Wall)
2. Position/Rotation of patient
3. Diaphragm
4. Clarity of Chostophrenic angles
5. Lung Parenchyma
6. Heart borders/Mediasteinum
7. Trachea
8. Presence of tubes, wires, and/or caths

AP View

Neonate Chest

Bony Structures
- Asymmetry
- Fractures
- Lesions
- Abnormal Curvature of spine

More about the bony structures
- Shoulder & Humorous should be visible on a CXR
- Is the patient rotated???
- Asymmetry of clavicles in A/P film
- Identify the medial border of the Scapula NOT a Pneumothorax

The Ribs:
- Note Obvious FX
- Asymmetry
- Trace rib from the vertebral origin and go anteriorly & inferiorly to the sternum
- Floating ribs don't articulate anteriorly (may need abdominal X-ray to see floating ribs)

The Sternum
- Poorly visualized on A/P films
- If suspected injury, refer to lateral film

Airways
- Trachea midline
- Tracheal deviation: toward or away
- Carina, mainstem bronchi
- Air bronchogram?

- Consolidation

Bronchogram

Lateral Neck
- Epiglottitis (Thumb)

Anterior Neck
- Croup (Steeple)

Epiglottitis

Right Lung
- Contains 3 Lobes
- Right upper (RUL)
- Right middle (RML)
- Right lower (RLL)
- Diaphragm is higher due to the position of the Liver (Under the 10th rib)

Left Lung
- Contains 2 Lobes
- Left upper (LUL)
- Left lower (LLL)

The Diaphragm
- Atelectasis
- Obesity
- Pleural effusion
- Emphysema
- Pregnancy
- Peritoneal fluid
- Bowel obstruction
- Tension Pneumothorax

Costo-phrenic angle
- Should have crisp edges on either side of the diaphragm
- Requires a minimum of 250cc of volume to disrupt clarity.

Pleural Effusions

Pneumonia
- Clinical History is imperative in Dx
- AKA:
- Consolidation
- Infiltrate
- Patchy white area
- Typically appear "Granular"

Kerley lines
- Kerley A: long, isolated, rarely seen. Upper and peripheral regions.
- Kerley B: stacked parallel lines laterally at bases.
- Kerley C: network of obscures normal branching and tapering of vessels to lower lungs.

Lung disease patterns
- Opacification:
- "White-out" consolidation or fluid filled alveoli
- Alveolar Filling
- Soft, "fluffy" infiltrate poorly demarcated
- "Ground glass"
- IRDS, ARDS, pulmonary edema, some pneumonia

ARDS

IRDS

Congestive Heart failure (CHF)
- Initially Identified Clinically
- Identified by PMHx
- Presence of Cardiomegaly
- Shaggy appearance around the heart border
- Kirly-B lines on film from interstitial edema
- Pleural effusions

Heart
- Cardiac shadow, aortic knob.
- Heart size < 1/2 width of film (PA)
- Ideally 1/3
- Cardiomegaly?
- Right heart border visible?
- If obscured, RML consolidation

Measuring Heart size

Cardiomegaly, CHF

Pulmonary angiography
- Radio-opaque dye injected into pulmonary arterial tree.
- Structures are visualized on a series of chest films.
- Definitive diagnosis of PE

Angiography

Endotracheal Tube Placement
- Locate the end of the ET tube
- Locate the carina
- Proper tube placement should be 3-4 cm above carina

Pneumothorax
- The absence of aeration throughout the lung field signifies the collapse of a lung
- Look for a black crescent over the apex of the lung.

Things that make you go hmmm

Chest Tubes
- Do not Place if there is one liter or more of blood loss.
- Apparent by entry into the pleural space
- Look for the presence of subcutaneous air around the chest tube.
- Look for improvement of lung inflation after insertion.

Wide Mediastinum

Head CT Scans

Once, a cat called X easily recognized by his black and white spots was a Ventriloquist. One day, he had a big break and masses came to see him.

The X:
- looks for symmetry in the film
- May account for asymmetrical appearances

Black spots:
- Infarct or old blood

White spots:
- acute blood

Ventriloquist:
- Ventricles, are they compressed or enlarged
- Sulci, are they too big or too small

Breaks:
- Are there any fractures?

Masses:
- Are they new or old?

OR.....The 3 "B's" of Trauma
- Bleeding
- Bruising
- Breaks

Types of bleeds...

Intraparenchymal
- White area within the tissue itself

Subarachnoid
- white are in sulci or the subarachnoid level of CT (star films)

Epidural
- white convex appearing lens

Subdural
- white moon or crescent shaped bleed

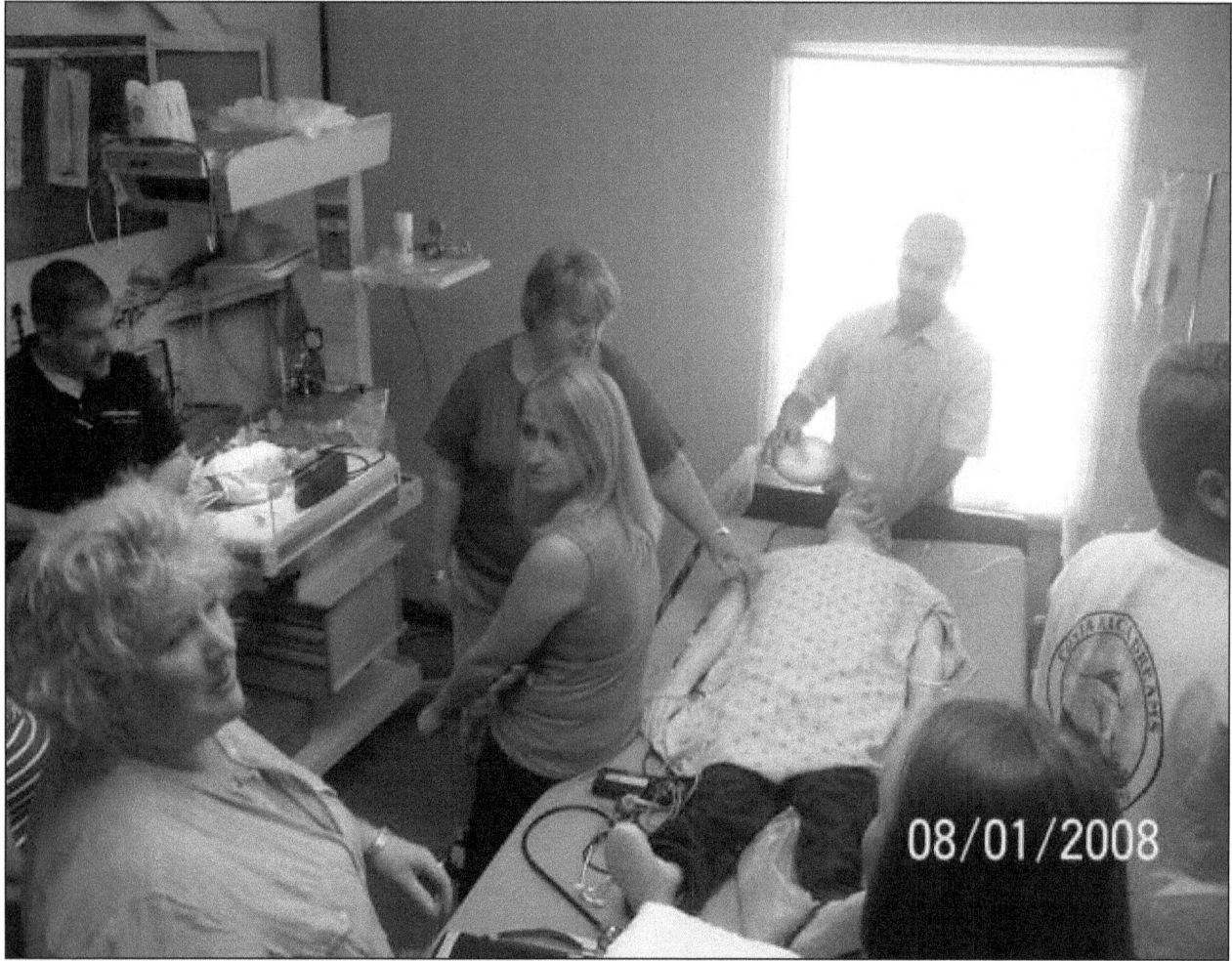

Advanced Cardiac Life Support (ACLS-EP) **Enhanced Provider** certification class
"Mega Code" (Accepted by CCEMT-P & FP-C for CEU's)

Toxicological Emergencies

The Basics

SCENE SAFETY!!!

- Remove/Decontaminate
- Primary Assessment/ ABC's
- Address life threats
- O2/ IV's
- Secondary Survey

ABC's of Toxicology
History:
a.

b.

c.

d.

e.

f.

g.

Accidental vs Intentional
- PMH
- Antidote available
- Psychosocial considerations

Antidote

Positioning

Gastric emptying/Lavage

Change Catabolism
Enhance Elimination

ABC's
- Positioning
- Phoenix Position- L Lateral Decubitus

Emesis Induction- Controversial
- No documented improvement in survival
- ABC's
- Lavage
- Controversial if ingestion is > 1 hour

Contraindications:
- CNS depression unless airway is protected

ABC's
- Charcoal-absorbs drugs and poisons
- May absorb antidote
- Enhance Elimination

Control seizures:

Flicker Effect

Antidotes
- Oxygen
- Narcan

Cyanide Antidote Kit
- CaCl or Gluconate
- Carbon Monoxide

Organophosphates
- Atropine
- 2 Pam Chloride
- Benzodiazepines (Flumazenil-administered with caution)
- Ca Channel Blocker

Weapons of Mass Destruction
- Nerve Agents

Salicylate Overdose
- ASA, OTC Cold meds, Pepto, Antacids
- Acute vs Chronic Ingestion

Signs/Symptoms:
- Tinnitus
- Coagulopathies
- Seizures
- HA
- Tachycardia/Tachypnea
- Confusion

Salicylate OD Tx

Acetaminophen OD

4 Stages

Stage 1
- 30 min-24 hours

Stage 2
- 24 – 48 hours
- RUQ Pain
- Oliguria - ATN

Stage 3
- 48 – 96 hours
- Peak Liver function abnormalities
- Fatalities d/t hepatic necrosis

Stage 4
- 4 days – 2 weeks
- Liver function returns to normal

Antidote – NAC-N-Acetylsysteine (Mucomyst)
- Give if known ingestion and if time of exposure known
- Charcoal will bind NAC

Accidental/Intentional

Chronic/Acute

Signs/Symptoms:
- N/V/D
- Photophobia
- Confusion/Restless
- Psychosis/Hallucinations

Cardiac Glycosides
- Digoxin, Oleander, Foxglove

Chronic Toxicity:
- Bradycardia & hypokalemia

Acute Toxicity:
- Tachycardia & hyperkalemia
- Blocks
- VT/VF

Treatments:
- ABC
- IV/O2
- Position
- Lavage-emesis is not recommended-produces vagal response.

TCA Overdose
- Elavil, Tofranil, Pamelor
- Well absorbed, difficult to remove
- Half life is 9 – 198 hours
- Chronic/Acute

Signs/Symptoms:
- HTN/ then hypotension
- Dry Mucous Membranes
- Arrhythmias
- Metabolic Acidosis

Treatment
- ABC's/O2/IV
- Position
- Cardiac monitor
- Prevent/Monitor conduction delays/arrhy.

Benzodiazepine OD
- Valium, Dalmane, Versed, Xanax
- Chronic/Acute

Intentional/Accidental

Signs/Symptoms:
- Dizzy
- Hypotension
- Drowsy
- Tachycardia
- HA
- Seizure
- Palpitations
- Coma
- Dry Mouth

Treatment
- ABC/IV/O2
- Position
- Charcoal/Lavage
- Flumazenil

Carbon Monoxide
Sources:
- Internal combustion engines
- Running machinery/poorly ventilated areas
- Poorly vented water heaters, furnaces and fireplaces
- Tobacco smokers
- Firefighters/Victims

CO Levels
- 10 – 20%
- 20 – 30%
- 30 – 50%
- 50 – 60%

HA, Nausea, impaired judgment!

Signs/Symptoms

- Initial Tachycardia/Tachypnea secondary to hypoxia then:
- Decreased HR
- Twitching
- A.Fib
- PaO2 decreased, O2 sat normal

Treatments
- Scene Safety
- Hi Flow O2

Concurrent cyanide poisoning

Ethylene Glycol / Methanol

Clinical Manifestations
- Profound Anion-gap
- Osmolar gap
- Nystagmus
- Depressed DTR's

- Stupor / coma / convulsions
- Myoclonic jerks
- Hypothermia / low grade fever
- Profound hypocalcemia

Treatment:
- IV Ethanol drip
- Fomepizole
- Thiamine
- Pyridoxine

Cocaine
- Avoid ß-blockers
- utilize alpha blockers for HTN

Hallucinogens
- Beware of violent tendencies

Narcotics
- Naloxone, titrated to arousal level and respirations

Toxin	Antidote	Notes
Carbon Monoxide	Oxygen	
Cyanide	Amyl & Na+ nitrate Na thiosulfate	
Organophosphates	Atropine, 2-pam	
Methemoglobinemia	Methylene Blue	
Anticholinergic	Physostigmine	
Coumadin	Vit- K+, FFP	
Heparin	Protamine sulfate	
ß & Ca++ blockers	Glucagon, Calcium	

Cyanide Poisoning
- Cyanogen chemical group
- Electroplating
- Ore extraction
- Production of combustion
- Seeds of cherries, apples, pears & apricots

Cyanide OD
- Cyanide combines and reacts with Ferric ions of the respiratory enzyme cytochrome oxidase to inhibit cellular oxygenation.
- HI Flow O2- displaces cyanide from cytochrome oxidase

Cyanide Antidote Kit-

- Converts Ferrous ions (Fe^{2+}) to Ferric ions (Fe^{3+}) forming methemoglobin. Cyanide has a greater affinity for iron in the Ferric state-combines with methemoglobin then the cyanide is released from the cytochrome oxidase and oxidase returns to its normal function in cellular respiration
- Methemoglobin does not carry O2
- Must convert Back to Hgb

Acid/Alkali

Acids:
- Hydrochloric,Sulfuric,Carbonic
- Tend not to penetrate deeply
- Tissue type necrosis
Alkalis:
- Lye, Chlorox, Ammonia

- Liquefaction necrosis
- Penetrate deeply

Scene Safety
- Dilute with milk or water
- Irrigate/Flush

Organophosphates
- SLUDGE Symptoms
    1. Salivation
    2. Lacrimation
    3. Urination
    4. Defecation
    5. Gastric Emptying
    6. Emesis
- Over 900 currently available
- Agricultural products, insecticides
- Interfere with breakdown of Ach at the NMJ – too much Ach
- Salivation
- Lacrimation
- Urination
- Defecation
- Gastro-intestinal complications
- Emesis
- Garlic odor

Treatments
- Scene Safety
- ABC's/IV/O2
- Position
- Lavage/Charcoal
- Atropine-blocks effects of Ach/dries secretions

Methamphetamine OD
Ingredients:
- Acetone
- Anhydrous Ammonia
- Red Phosphorus

Snort/Smoke/Inject
- Concurrent Ingestion
- Hi Flow O2

Remove all clothing! SCENE SAFETY

RSI / Advanced Airway

Definition
"The administration of potent induction agents followed by rapidly acting neuromuscular blockers to induce unconsciousness and motor paralysis for tracheal intubation."

Ron M Walls, MD
Manual of Emergency Airway Management

Indications
- Failure of airway maintenance and/or protection
- Failure of ventilation or oxygenation
- Anticipated clinical course

Contraindications
- Probable inability to intubate / ventilate
- Relative contraindications of a specific agent

Procedure and Technique
- Preparation
- Preoxygenation
- Pretreatment
- Paralysis
- Protection & Positioning
- Placement with proof
- Postintubation management

RSI: Preparation
- Assess carefully for indicators of a difficult airway!
- Have a fall back plan.
- Cardiac monitoring, SaO2, NIBP in place, positioned.
- Minimum of one patent IV line.
- Patient should be optimally positioned.
- All equipment should be tested.
- Post intubation and back up equipment

RSI: Preoxygenation
- "The establishment of an oxygen reservoir within the lungs and body tissues to permit several minutes of apnea without arterial oxygen desaturation."
- Essential to the optimal "no bagging" RSI.
- Administration of 100% O2 for 5 minutes will replace the primarily nitrogenous residual lung volume with 02.

Time to desaturation is patient dependent.
- A healthy 70kg adult can maintain SaO2 of >90% for approximately 8 minutes with appropriate preoxygenation.
- A 130kg obese adult will desaturate to <90% SaO2 in approximately 3 minutes.
- A 10kg child will desaturate to <90% in less than 4 minutes.
- Desaturation times from 90% to 0%
- Accelerate as the saturation grows lower.
- Approx 120 seconds 90 – 0 in a healthy adult 45 seconds in a child.

RSI: Pretreatment
The administration of drugs to mitigate adverse effects associated with the intubation.

"LOAD"
- Lidocaine 1.5mg/kg >3 minutes prior to laryngoscopy blunts ICP rise and decreases bronchospastic reactivity in asthma patients.
- Opioid: Fentanyl in particular will blunt the sympathetic response to laryngoscopy.
- Atropine: Children less than 10 yrs 0.02mg/kg
- Defasciculating agents

RSI: Paralysis with induction

Rapidly acting induction agent
- Etomidate, versed, ketamine, etc.

Rapidly acting paralytic agent
- Succinylcholine, etc.

Technique is based on rapid loss of consciousness, rapid neuromuscular blockade, and a brief period of apnea without interposed bag mask ventilations before tracheal intubation.

RSI: Protection and Positioning
- Sellick's maneuver to prevent passive regurgitation.

BURP
- Backwards
- Upwards
- Rightwards
- Pressure

RSI: Placement and Proof
- Positive cord visualization
- Minimum of 3 confirmation methods with at least 1 type of EtCo2 detection.
- + BBS / -epigastric sounds
- + Tube fogging
- + SaO2 rise
- + EtCo2 waveform
- + EtCo2 colormetric change
- Esophageal detection device

RSI: Postintubation Management
- ETT must be secured
- Commercial device is preferable for transport
- Long term sedation
- Long term paralysis

Pain management
- Continuous vital sign monitoring
- ETT placement confirmation with EACH move!

RSI: Post intubation Management
- Hypotension in the post RSI period:
- Tension pneumothorax:
- Change from negative to positive pressure ventilations
- Over aggressive ventilations
- Decreased venous return filling
- Positive pressure ventilations inhibiting cardiac

Induction agent side effects
- Tachycardia in the post RSI period:
- Need for additional sedation
- Change in patient condition

Timing
- RSI requires knowledge not only of the required steps, but of the appropriate timing.
- 5 minutes: Preoxygenation or 8 vital breaths
- 3 minutes: Premedication
- 2 minutes: Sedation
- 1 minute: Paralytic

INTUBATION
- +1 minute: Proof and post intubation care

RSI Medications: LOAD
Lidocaine:
- 1.5mg/kg IV 3 minutes prior to intubation.
- Suppresses cough reflex
- Suppresses bronchospasm
- Blunts ICP rise
- Increases the depth of anesthesia
- Decreases cerebral metabolic oxygen demand
- Decreases cerebral blood flow

Opioid:
- Fentanyl 3 -6 mcg/kg
- No direct effect in ICP
- No histamine release
- Partial attenuating effect of sympathetic response to laryngoscopy
- Little to no cardiac or vasodilating effect
- Given over 30 seconds to avoid thoracic muscle wall rigidity.
- Remember, it is not reversible with Naloxone

Atropine:
- 0.02mg/kg 3 minutes prior to intubation
- Primarily given as pretreatment in children
- 10 yrs or younger.
- Any adult who receives a 2nd dose of Succinylcholine should have 0.5 – 2 mg readily available.

Defasciculating agents
- Non-depolarizing agents
- Norcuron, Rocuronium, Pavulon, etc.
- Normal defasciculating dose is 10% of the paralyzing dose.
- Mitigates potential ICP rises
- Blunts fasciculation's
- Pavulon can cause tachycardia, potentially precluding the need for atropine in children

RSI Medications: Sedation

Induction

Three categories:
- Ultra-short acting barbiturates
- Benzodiazepines
- Miscellaneous agents

Ultra short acting barbiturates

Thiopental (Pentothal) 3 – 5 mg/kg adult
- Cerebroprotective
- Decreases cerebral metabolic O2 demand
- Potent vasodilator
- Anticonvulsant properties

Methohexital (Brevital) 1.5 mg/kg adult
- 3 – 4 times more potent than Pentothal
- Potent vasodilator
- Both are rarely used in today's environment

Benzodiazepines

Midazolam (Versed) 0.2 mg / kg IV
- Amnestic properties
- Dose related hypotension

Diazepam (Valium) 0.3 – 0.5 mg / kg IV
- Slower acting
- Amnestic properties not as good as Versed

Lorazepam (Ativan) 0.1 mg / kg IV
- Slowest acting and longest lasting
- Amnestic properties not as good as Versed

Miscellaneous Agents

Etomidate (Amidate) 0.2 – 0.3 mg / kg
- Hypnotic with minimal cardiac or respiratory effects.
- Onset of 15 – 30 seconds, with recovery in 7 min
- Can cause projecting vomiting with rapid IV push

Ketamine (Ketalar) 1 – 2 mg / kg IV
- Dose dependent myocardial depression and increased ICP
- Considered a good choice in asthmatics
- Onset of 15 – 30 seconds, recovery in 15 min.

Propofol (Diprivan) 0.5 – 1.2 mg / kg IV
- Increases ICP
- Myocardial depressant
- Vasodilating effect
- Excellent sedation and hypnotic properties

RSI Medications: Paralytics

Neuromuscular blocking agents fall into two categories:

Noncompetitive Depolarizing NMBA
- Succinylcholine (SCh) 1 – 1.5 mg / kg

Competitive Nondepolarizing NMBA
- Vecuronium (Norcuron) 0.1 – 0.25 mg / kg
- Rocuronium 0.6 – 1.2 mg / kg
- Pancuronium (Pavulon) 0.08 mg / kg
- Rapacuronium (Rapalon) 1.5 mg / kg

Noncompetitive Depolarizing NMBA
Succinylcholine (SCh) 1 – 1.5 mg / kg
- Binds tightly to acetylcholine receptors
- Resulting depolarization manifests in fasciculation's
- Paralysis in 45 – 60 seconds
- Recovery in 5 – 15 minutes in adults
- Dose in pediatrics is 2 mg / kg
- Dose in newborn is 3 mg / kg
- Little issue with giving too much SCh
- Significant side effects to giving too little SCh

Side effects:
- Fasciculation's increase ICP, IOP, IGP.
- Defasciculating with Nondepolarizing agent, or with 0.15 mg / kg of SCh.
- Hyperkalemia: increase of 0.5 mEq/L serum potassium.
- No SCh use in hyperkalemic patients or in significant burn or crush injury victims >24 hrs post injury.
- Bradycardia in pediatrics or in repeat adult dosing
- Prolonged paralysis due to pseudocholinesterase deficiency typical in organophosphate poisoning or cocaine overdose.

Malignant hyperthermia (MH) is a genetic muscle membrane abnormality
- 60% mortality
- Muscular rigidity, autonomic instability, hypoxia, hypotension, severe lactic acidosis, hyperkalemia, myoglobinemia, DIC are early signs
- Elevations in temperature are a late manifestation

Treatment is with Dantrolene Sodium (Dantrium)
- 2.5 mg / kg repeated q 5 minutes until symptoms subside to a max dose of 10 mg / kg.
- Full paralysis with a nondepolarizing agent will prevent SCh induced MH.

Trismus / Masseter muscle rigidity
- Occurs most commonly in children
- Usually transient, but continued rigidity can be a sign of MH
- Pretreatment with a defasciculating agent will not prevent masseter muscle rigidity.
- Full dose of a nondepolarizing agent will cause relaxation.

Contraindications:
- Family history of MH is an ABSOLUTE contraindication
- History of any progressive muscular disease
- Probable inability to intubate / ventilate
- Burns or crush injury > 24 hours since incident
- Metabolic hyperkalemia
- Increased ICP is a relative contraindication

Compete with and block action of acetylcholine at the neuromuscular junction receptor site

Vecuronium (Norcuron) 0.1 – 0.25 mg / kg
- Does not promote tachycardia
- Onset can be decreased by increasing the dose, but duration will increase as well.
- Onset in 90 – 120 seconds
- Duration of 30 – 45 minutes

Can be reversed with administration of
- Neostygmine (Prostygmine) 0.06 – 0.08 mg /kg

Competitive Nondepolarizing NMBA

Rocuronium 0.6 – 1.2 mg / kg
- More rapid onset than Norcuron
- Onset similar to SCh
- Duration of approximately 30 minutes

Can be reversed with administration of
- Neostygmine (Prostygmine) 0.06 – 0.08 mg /kg

Pancuronium (Pavulon) 0.08 mg / kg
- Tends to produce tachycardia
- Average of 3+ minutes for paralysis
- Recovery time of approximately 90 minutes

Can be reversed with administration of
- Neostygmine (Prostygmine) 0.06 – 0.08 mg /kg

Rapacuronium (Rapalon) 1.5 mg / kg
- Rapid onset identical to SCh
- Potential for tachycardia, hypotension, and significant bronchospasm.
- Onset of > 1 minute
- Recovery in approximately 10 – 20 minutes
- Currently pulled off the market

Can be reversed with administration of
- Neostygmine (Prostygmine) 0.06 – 0.08 mg /kg

RSI: The Decision to Intubate

Indications:
- Failure of airway maintenance or protection
- Failure of ventilation or oxygenation
- Anticipated clinical course

Failure of airway maintenance or protection
- Any patient who requires establishment of an airway requires protection of that airway.
- A patient with a spontaneously patent airway may not be able to protect themselves from aspiration.
- Ability to swallow may be a better indicator of need for protection than absence or presence of a gag reflex.
- Potential for airway obstruction / closure.

Failure of ventilation or oxygenation
- Lack of adequate ventilation
- Inability to adequately move air
- Fatigue
- Pain
- Obstructive process
- Injury
- Inability to oxygenate
- Dead space ventilation (asthmatics)

- Injury or metabolic process

Anticipated clinical course
- Presence of shock
- Potential for airway obstruction
- Potential for aspiration
- Inability / unwillingness to ventilate adequately
- Combative behavior
- Transport / Treatment considerations

RSI: Overall Airway Management

RSI: Overview
- Solid, facts-based decision making process
- Back up plans and tools in place
- Seven P's of RSI

Solid knowledge of medications
- Proactive approach to changes in
- Protective approach to tube confirmation and continued patency patient status

RSI: Case 1
- 88 yr old male, was plowing his field with a large tractor when he fell off the machine, causing its large rear tire to roll over his left chest wall, and the plow to partially amputate both legs below the knees.

Issues:
- Temperature of 35 degrees F
- Patient has been in the field for > 1 hour before being found
- Due to rural location, you arrive with volunteer FD first responders, no EMS on scene.

Information:
- Age 88 yrs
- Weight 65 kgs
- Hx of MI nine yrs ago w/ stent placement, HTN
- Pt is on a beta-blocker for HTN
- No allergies
- GCS 13 (4/4/5)
- HR 86 A-fib on monitor; BP 110 / 76, Resps. 10/long, slow & shallow, SaO2 86% on room air.

GCS Chart
- Action plan?
- Medications?
- Doses?
- Order?
- Problems?

RSI: Case 2
- 3 year old female run over by a go-cart at approx 30 mph. Wheels ran across her chest cavity w/ obvious deformity.

Issues:
- EMS meets you at an LZ in a football field, and runs towards the helicopter as you land. Ambulance is >150 yards distant. A large crowd from the adjoining elementary school gathers outside to watch.

Information:
- Pt is met outside the rotor disk, currently immobilized with Pedi board, c-collar, tape. NRB is in place at 15 LPM O2. IV NS 22 ga. KVO in left foot.
- Pt weighs 17 kgs
- No family present to provide further information
- GCS: 6 (1/1/4)
- HR 145 ST on monitor, B/P 68/42, Resps 36 shallow and labored, SaO2 88% with O2 in place

GCS Chart

- Action plan?
- Medications?
- Doses?
- Order?
- Problems?

RSI: Case 3
- Responding to a rural ER for a MVC patient with confirmed C1 / C2 fracture and possible closed head injury.

Issues:
- Patient is 140 kgs
- Pt is in a c-collar, but not on LSB
- Pt is heavily bearded
- Information:
- O2 is in place NC at 4 lpm with SaO2 of 90%
- IV x 2 18 ga. bilateral ACs

- GCS of 13 (3/4/6)
- HR 132 ST on monitor, B/P 172/98, resps 26/rapid

GCS Chart

- Action plan?
- Medications?
- Doses?
- Order?
- Problems?

Mallampati score 1-4

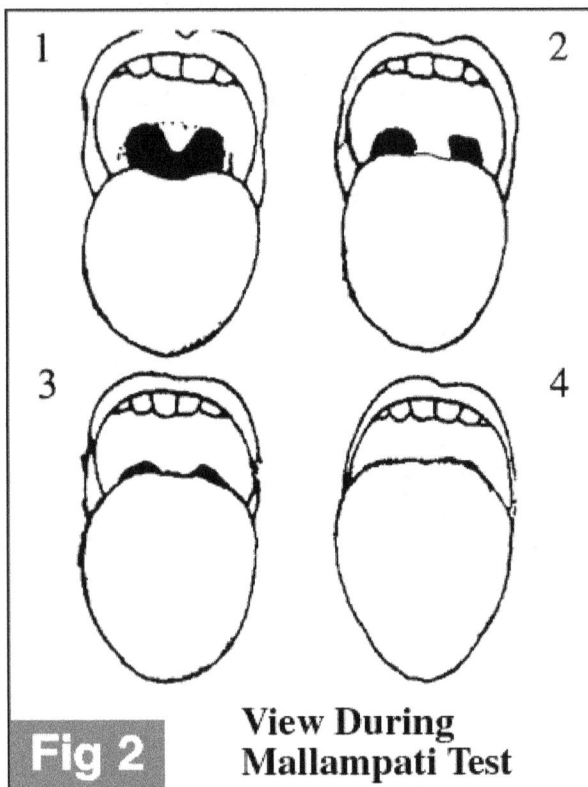

View During
Mallampati Test

Cricothyrotomy

A kit or device that is intended to establish a surgical airway without resorting to formal cricothyrotomy

2 methods
- Seldinger
- Direct percutaneous placement

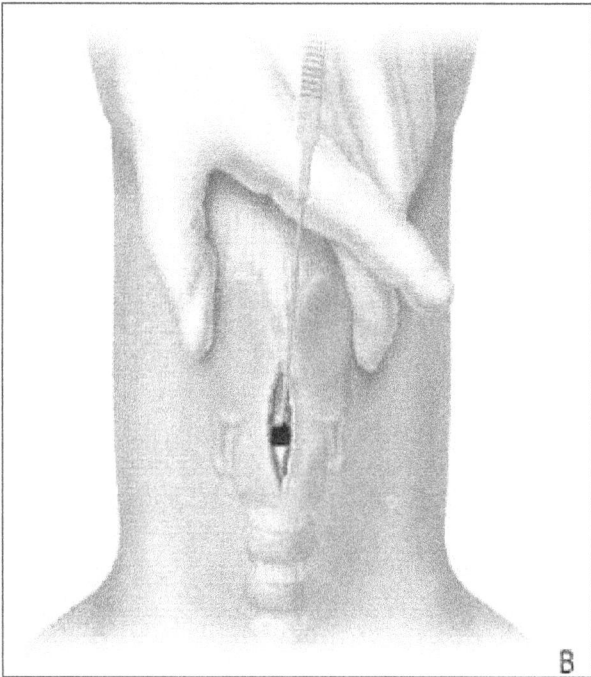

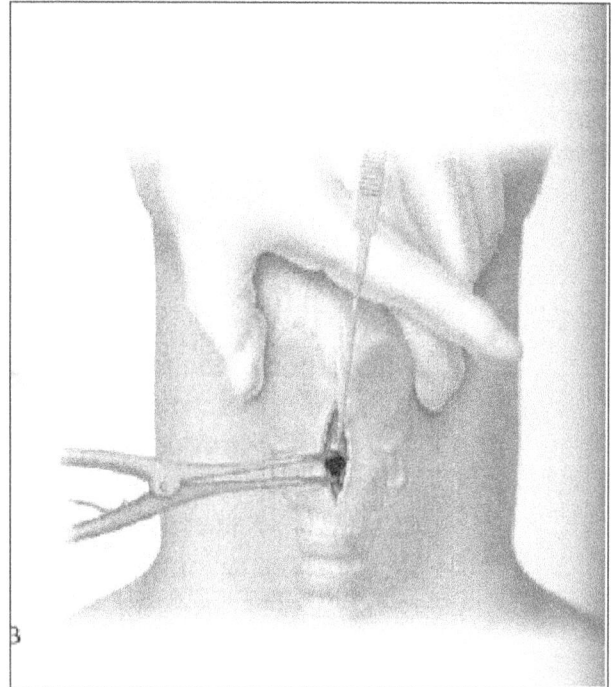

Indications
- Can't oxygenate
- Can't ventilate

Medical Emergencies

Diabetic Ketoacidosis (DKA)

- Insufficient insulin or cell's inability to use insulin
- Hyperglycemia causes osmotic diuresis, → glycosuria, dehydration, and electrolyte imbalances
- Glucose elevated 300-800mg/dL
- Excessive fatty acids enter liver leading to ketoacidosis

Treatment:

- Treat cause: infx, noncompliance with diet/meds, drug interaction
- Correct fluid volume deficit
- Insulin therapy: initial bolus, then slow IV infusion with Q1H BG √s
- When BG < 250 mg/dL, fluids should be changed from NS to D5NS
- Correct electrolyte and acid/base imbalances
- NaBicarb for severe acidosis only

Hyperglycemic Hyperosmolar State (formerly HHNK /HHNK)

- Hyperglycemia with profound Hydration without ketosis
- In HHS coma, sufficient insulin is released, but not enough to prevent hyperglycemia
- Hyperglycemia causes osmotic diuresis → glycosuria, dehydration, and electrolyte imbalances
- 3 Major Signs: Elevated blood glucose > 1000mg/dL and plasma hyperosmolarity > 350mOsm/kg and extremely elevated Hct

Treatment:

- Correct fluid balance, watch for seizures in hyperosmolar dehydration
- IV Insulin therapy with Q1H BG √s
- Correct electrolyte imbalances

Acute Abdomen/GI Hemorrhage
Don't forget ABCs...

- OGT/NGT, poss. gastric lavage when indicated
- Blood transfusion/fluid replacement
- Vasopressin gtt 0.2-0.6 units/min, consider NTG gtt to prevent coronary artery constriction/CP
- Sandostatin (↓upper GI mobility ↓gastric acid secretion) @ 25mcg/hr
- Differentiate Lower/Upper hemorrhage:
  - Upper—hematemesis, melena, epigastric pain/burning, hx NSAIDs/ulcers
  - Lower—hematochezia, occult blood, lower abd pain, diarrhea, cramping

Syndrome of Inappropriate Anti-Diuretic Hormone (SIADH)

- Increased secretion of ADH or increased renal responsiveness to ADH
- Impaired renal excretion of water → oliguria, <u>high</u> urine specific gravity, water intoxication and h<u>yponatremia</u>

Etiology:

- Neurogenic (↑ prod. or release of ADH) – CNS trauma/ Lung Dx.
- Ectopic (prod. of ADH like subst.) – Oat cell carcinoma,
- Nephrogenic (agents that ↑ or enhance ADH effects) – Tricyclics, sedatives, narcotics, diuretics
- Symptoms: personality Δ, HA, ↓ mentation, lethargy, N/V/D, ↓ tendon reflexes, seizures
- Serum Na less than 120 meq/L, Urine spec. gravity > 1.030, Urine osmolality > 1200 mOsm/L

Treatment:

- Treat the cause: Dilantin/Lithium inhibits action of ADH on renal tubules, especially in ectopic ADH
- Correct fluid volume excess and electrolyte imbalances

- Hypertonic (3%) saline 1-2 ml/kg/hr until serum Na reaches 125 meq/L

Diabetes Insipidus

- Deficiency of ADH or inadequate renal tubule response to ADH
- Impaired renal conservation of water → polyuria, <u>low</u> urine specific gravity, dehydration and <u>hypernatremia</u>
- Etiology: Neurogenic/Central (defect in release of ADH) --closed head injuries/pituitary tumors) or Nephrogenic (defect in renal tubular response) – renal dx or drugs— Dilantin, caffeine, Norepi
- Symptoms: thirst, weakness, poor skin turgor, irritability, confusion, seizures
- Serum Na > 145 meq/dL, Urine spec. gravity 1.005, Urine osmolality < 200 mOsm/L

Treatment:
- Pitressin, DDAVP, aqueous vasopressin or hypophysectomy for Neurogenic DI
- Thiazide diuretics and Na restriction for Nephrogenic DI
- Correct fluid deficit with hypotonic solutions--½ NS or D5W
- Replace electrolyte imbalances

Grave's Disease / Thyrotoxicosis / Thyroid Storm)

- Formation of auto antibodies that bind to TSH receptor and stimulate the gland to hyperfunction
- S/S: heat intolerance, fatigue, exophthalmoses, irritability, palpitations, tachycardia
- Confirmed by increased T3 and T4 levels
- Tx: Medical—Propylthiouracil (PTU) or Methimazole & glucocorticoids
      Surgical—thyroidectomy
- Plasmapheresis has been used to treat thyroid storm in adults

Hypothyroidism/Myxedema Coma

- Occurs from untreated or inadequately treated hypothyroidism OR precipitated by infx, CVA, MI, acute trauma, excessive hydration, or admin. of sedatives
- Majority of cases occur in winter months
- S/S: hypothermia, hypoventilation, hyponatremia, hyporeflexia, hypotension, bradycardia…continuous seizures when death is imminent
- Tx: IV Levothyroxine 0.2-0.5mg, Synthroid, warming blankets, Hydrocortisone to treat poss. adrenocortical insufficiency

Acute Adrenal Insufficiency/Addison's Disease

- Most common cause is related to autoimmune destruction of adrenals (primary Addison's) or less often (2∘ Addison's) ; trauma, surgical procedure, tumors/ca
- S/S: anorexia, N/V/D, fatigue, irritability , melanin pigmentation
- Tx: IV Glucocorticoids / Mineralocorticoids, Monitor K & Na

Hypercortisolism/Cushing's Syndrome

- Marked increase of production of mineralocorticoids, glucocorticoids, and androgen steroids r/t adrenal & pituitary tumors

S/S:
- Hyperglycemia, acidosis, ↑cortisol level, gynecomastia, facial hair, acne, "Moon face", "buffalo hump", mood swings, insomnia

Tx:
- Surgical removal of tumor, steroid replacement, monitor K and Na

Pancreatitis
- Causes: ETOH abuse, obstruction of bile duct, drug induced (diuretics/steroids/antibx), trauma/hemorrhage

- S/S: abd. pain, ↑amylase/lipase, Grey-Turner's/Cullen's sign (trauma), Chovostek's/ Trousseau's (hypocalcaemia), Δ breath sounds: Atelectasis/pleural effusions/ARDS
- Tx: NPO/OGT, aggressive fluid replacement, pain management, Somatostatin, √ lytes, BG

Rhabdomyolysis

- Skeletal muscle necrosis assoc. with:
  - Trauma/crush injuries, ETOH, cocaine/heroin, hyperthermia, infx, drugs—statins
- Symptoms: Confusion, febrile, CK level is often > 10,000-100,000 U/L, Myoglobinuria (dark, tea-colored urine)
- Urine hemoglobin positive in approx. 50% of patients
- Increased serum Creatinine as renal failure develops
- Differential diagnosis: In trauma pts, localized pain may resemble s/s of deep venous thrombosis.

Treatment:

- Treat underlying cause: fasciotomy, cooling blankets, IV Antibx
- IV Fluids NS or D5 1/3NS @ 500cc/hr
- Consider NaBicarb to keep urine acidity > 7.5
- Dialysis is indicated in renal failure if the patient is anuric and diuresis is not induced with rehydration.

Disseminated Intravascular Coagulation (DIC)

- Syndromes characterized by thrombus formation and hemorrhage 2° to overstim. of coagulation process → decrease of clotting factors and platelets
- Etiology is always secondary: Vascular disorder, infx/sepsis, Sickle cell crisis, trauma, obstetric complications, ARDS, IABP, Snake bites, Aspirin poisoning
- Triggered by intrinsic and extrinsic coagulation system activation and red cell

or platelet injury, clotting causes ischemia and tissue and organ necrosis, bleeding causes loss of Hgb and oxygen carrying capacity, fibrinolysis causes destruction of once stable clots and further bleeding

- *Symptoms*: CP, dyspnea, abd. Pain, ΔLOC, ↓ UO, Petechiae, occult bleeding
- *Labs*: platelet count < 150,000/mm³, PT > 40sec, APTT > 70sec, Fibrinogen < 200mg/dL, D-dimer elevated/+
- *Treatment*: Treat underlying condition (easier said than done), Heparin therapy, whole blood, FFP, Vit. K, folic acid, Cryoprecipitate

Here's your Sign...

1. Brudzinski's

2. Kernig's

3. Cullen's

4. Grey Turner's

5. Kehr's

6. Murphy's

7. Levine's

Obstetric Emergencies

What Happens When You're Pregnant?

- Cardiovascular
- Respiratory
- Musculoskeletal
- Neurological
- Genitourinary
- Gastrointestinal
- Reproductive
- Hematological

Cardiovascular
- Increased plasma volume-45%
- Increased cardiac output-40%
- Tachycardia-Due to increase in blood volume-increases 15%
- Hypotension-Systolic decreases 3-5 torr
- Diastolic decreases 5-10 torr

Respiratory
- Chest cavity expands laterally
- Increased oxygen consumption
- Decreased arterial pCO2 and serum bicarbonate

Musculoskeletal
- Displacement of abdominal viscera-altered probability of injury and altered pain referral pattern.
- Ligamental tissue increasingly elastic
- Widened pelvic girth
- Center of gravity is different
- Pelvic venous congestion-increased risk of hemorrhage due to injury.

Neurological
- Tingling
- My Feet are Killing Me!!!!

Genitourinary
- Increased blood flow
- Increased urine output
- Increased water retention
- Decreased specific gravity

- Dilation of ureters and urethra with bladder displacement
- Increased risk for urinary tract infections

Gastrointestinal
- Decreased peristaltic action
- Decreased gastro esophageal sphincter competency
- Increased gastric reflux
- Greater chance for aspiration

Reproductive
- Uterine enlargement and displacement of bladder-increased vulnerability to injury.
- Increased pelvic vascularity-potential for significant blood loss.

Hematological
- Increased clotting factors
- Decreased fibrinolytic activity
- Hypercoagulability

Obstetrical Conditions

1. Placenta Previa
2. Prolapsed Cord
3. Embolism
4. Abruptio Placenta
5. Abruptio Placenta-Causes
6. Placenta Previa
7. Placenta Previa-Causes
8. Multiparity
9. Edometritis
10. Congenital malformations/fetal presentation

Differences (Know these!)

Abruption Placenta = Always Painful
- Dark uncoagulated blood
- Painful, shearing like pain

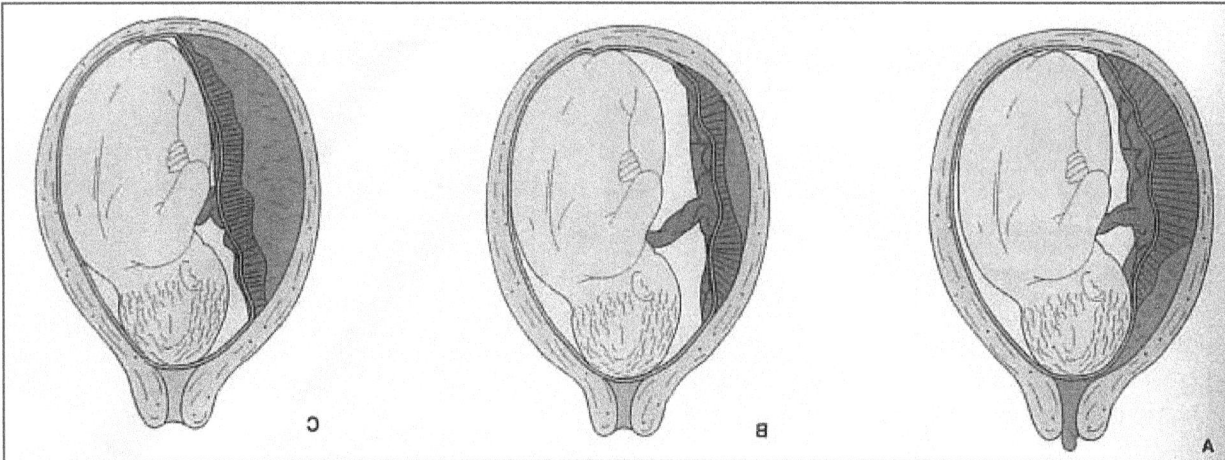

Placenta Previa = Painless Privates
- Bright red bleeding
- Painless

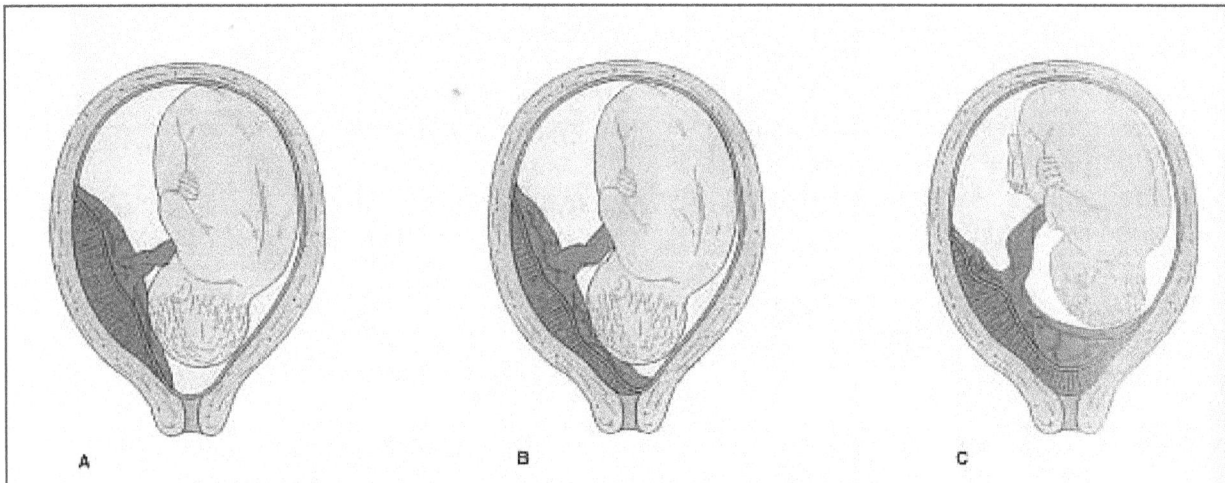

Treatment
- Monitor maternal VS and urine output
- Loss of 1800 cc will cause hypotension
- IV rates to maintain BP, utero-
- placental perfusion and urinary output > 30 cc/hr
- Blood replacement on a case by case basis with PRBC or FFP

Uterine Rupture
- Associated with direct abdominal impact
- Rupture of the uterus-usually along former C-section scar, operation scar
- Obstetrical emergency
- Severe, sudden, shearing pain during strong contraction, absent FHT, hard mass next to fetus, palpation of fetal parts, minimal external bleeding

Prolapsed Cord
Treatment
- Place mommy in knee/chest position or elevate hips.
- Two gloved fingers into vagina with cord between fingers, upward pressure until pulsation of cord is felt.
- Once your hand is there, it stays there and so do you!!!!!

Uterine Inversion
- Protrusion of uterine fundus beyond cervix, profuse bleeding, sudden, severe, lower abdominal pain.
- Occurs after contraction
- Associated with cough, sneeze, improper fundal massage or traction on cord
- Embolism
- Pulmonary

Amniotic Fluid

Pelvic Fracture
- Two-point displacement
- Retroperitoneal bleeding
- Mast-NO inflation of abdominal compartment

Pre-term Labor/PROM
- Pre-term Labor-Regular contractions between 20-36 weeks.

- PROM-Spontaneous rupture of amniotic membrane before the onset of labor occurring prior to 37 weeks.

Treatment
- IV Fluid Bolus
- Tocolytics
- MgSo4-Smooth muscle relaxer
- 4-6 grams/250 cc NS over 20-30 minutes
- IVPB-1-3 grams/hour
- Antidote-calcium gluconate-1 gram 10% over 3 minutes
- Brethine-beta-sympathomimetic 0.25 mg SQ q 4 hours
- May give q 20 minutes X 3 doses
- Antidote-Inderal 0.5 mg slow IVP (You will see this one again) or Verapamil HCL 5-10 mg slow IVP

Hypertensive Disorders

PIH (Pregnancy Induced Hypertension)

- Superimposed Hypertension
- HELLP Syndrome (Hypertension, Elevated Liver Enzymes and Low Platelets)

Treatment

- IV fluids total-100 cc/hr or less
- Anti-hypertensive's
- Apresoline-5 mg IV followed by 10 mg q 20 minutes until diastolic < 110
- Labetalol-10 mg IV followed by 20, 40, 80 mg q 10 minutes until total 300 mg or diastolic < 110
- Nifedipine 10 mg/min doubled q 5 min until diastolic < 110

Remember........
- Think of eclampsia as a possible cause of injury in the pregnant trauma patient with altered mental status, seizures, or hypertension!

Etiology of Major Trauma

Trauma
- Blunt-MVC most common
- Leading cause of maternal death-head injury
- Leading cause of fetal death-maternal death
- Leading cause of fetal death if mother lives- abruption
- Labor produced by release of arachidonic acid and/or thromboplastin

Motor Vehicle Crashes

Falls
- Forward momentum due to gravid uterus
- How did she land?
- What did she fall on?

Penetrating Trauma
- Gunshot Wounds-poor prognosis
- Stab Wounds-better prognosis but still bad
- Perinatal mortality is 41-71%
- Maternal mortality is < 5%

Blunt Trauma
- Leading cause of non-obstetrical related fetal death during pregnancy.
- Blows to Abdomen-secondary to MVC
- Falls and assaults are next in line

Burns
- Enhanced risk for volume deficit and generalized hypoxia

Parkland formula:
- 4cc X weight (kg.) X BSA of burn
- Total in first 24 hours, ½ in first 8 hours

Assess, Don't Molest!!!!

- Oxygen by NRBM or Endotracheal tube
- Full Spinal Restriction
- Deflection of uterus off the inferior vena cava and aorta
- Crystalloid resuscitation-Aggressive
- Secondary survey-
- Head to toe
- Fetal assessment-Doppler every 15 min, during and after contractions
- Labor Assessment

- Cervical dilation
- Contraction status

How do we transport mommy?
- Don't let Mommy move!
- Roll left (or right)
- Everybody gets O's!
- Fluids are a good thing!
- We don't need lights and sirens!!!!
- Left or Right?

Probability of Fetal Death
- Fetal Survival vs. Gestational Age

Interventions for Fetal Distress
- Consider Delivery
- Continual Fetal Monitoring
- Postmortem C-Section

Gynecological Emergencies

Previous OB History
- Previous vaginal or C-section? VBACS?
- Location and extent of abd scars
- Were there any complications w/ her or baby?
- Preterm deliveries? Gestational age?
- Outcome?
- Previous spontaneous or elective abortions? D&C required?
- # of children currently? Weight & sex?
- Length of last labor?

Pertaining to current pregnancy
- Contractions? When did they begin? Change in intensity or frequency?
- Accompanying back, pelvic or rectal pain?
- Bleeding or "bloody show"? (# of towels)
- Was bleeding painless or associated w/ contractions or abd pain?
- Blood bright red or dark?
- Was there mucus in the blood?
- When did it begin?
- Activity prior to bleeding?

Pertaining to current pregnancy
- Bag of H2O's ruptured? Gush or trickle? (Small leakage may be confused w/ urinary incontinence)
- Leakage of amniotic fluid? Time it occurred?
- Color, meconium-stained, dark (presence of blood) or clear?
- Odors associated? Is the chux wet or pooling w/ fluid?
- Patient smoke or drink ETOH? Frequency and last time used?
- Adequate weight gain? Malnourished or obese?
- Prenatal care? Δ in fetal activity? Meds? Any probs w/ this pregnancy?
- Any diagnostic tests done?

Management
The best way to determine fetal well being...

Indicates adequately oxygenated fetus

PID
Ruptured Ovarian Cyst
- Spontaneous, may be associated with abdominal injury, intercourse, or exercise
- Thin walled fluid filled sac, benign
- Develops on ovary
- Bleeding is rare

Cystitis
- Frequent infection of bladder and ureters
- May lead to pyelonephritis
- Suprapubic tenderness, urinary frequency, dysuria, hematuria

Mittelschmerz
- Midway into menstrual cycle
- Pain at time of ovulation possibly due to follicular leakage/bleeding during ovulation
- Ovaries and follicles affected
- Unilateral lower quadrant abdominal pain, low grade fever

Endometriosis
- Average age-late 30's in women who defer pregnancy
- Growth of endometrial tissue outside of uterus
- Fallopian tubes, pelvic organs, bowel, bladder, ligaments
- Painful intercourse, painful menstruation, painful BM's
- Sever pain following intercourse and BM's

Ectopic Pregnancy
- Always consider with female of reproductive age
- Possible life threatening if rupture occurs
- Possible vaginal bleeding

Vaginal Bleeding

Placenta previa / placenta abruption (know these)

PID
- Labor
- Vaginal Bleeding
- May be life-threatening
- Check for impending shock, orthostatic VS

Traumatic Abdominal Pain
- Straddle injuries, blows to perineum, blunt force to lower abdomen, foreign bodies in vagina, abortion attempts, soft tissue injury
- Severe bleeding, organ rupture, hypovolemic shock

Sexual Assault
- Do not ask questions that will cause quilt feelings.
- She did not deserve it.
- Do not inquire about sexual history or practices.
- Sexual Assault
- Examine genitalia only if necessary
- Explain all procedures
- Avoid touching without permission
- Psychological support
- Safety
- Do not force patient to talk
- Do not use invasive procedures
- Preserve evidence
- Non-judgmental
- Female personnel if possible
- Confidentiality is critical

Amniotic Fluid Embolism
- Occurs when amniotic fluid gains access to maternal circulation during labor and delivery
- In addition to amniotic fluid, lanugo hairs, fetal squamous cells, bile, fat, and mucin may embolize
- Cause of deaths in 10% of maternal deaths
- It is rare with a mortality rate of 90%
- Most often misdiagnosed, due to vague symptoms, and missed on autopsy.

Factors associated are:
- Uterine rupture, C-section, and use of uterine stimulants to induce labor (produce hypertonic contractions)

Amniotic Fluid Embolism
- Risk Factor include:
- Placenta previa
- Intrauterine fetal death
- multiparity
- Knee-chest position
- DIC also...although pathway is unclear

Breech delivery
- Can be felt through the cervix during an exam
- Can be seen on ultrasound
- At, or near term risk is 3-4%, however before 34 weeks gestation, incidence is higher
- Does not affect transport unless in labor, and membranes are ruptured
- Labor is typically slower when breech
- Can palpate cord once legs and buttocks are delivered
- Gentle steady traction downwards once hairline is visable. The upwards traction holding feet high.
- Airway equip readily available

Pre-Eclampsia / Eclampsia
- Can occur pre, during, or postpartum
- S/S include:
- Headache, visual disturbances, apprehension, anxiety, hyper-reflexia
- Seizures 3rd trimester gestation
- Coma can ensue

Treatment:

HELLP Syndrome
Associated complications of pre-eclampsia

H -

E -

L -

LP -

PIH

Assessment:
The "big three"

1.

2.

3.

Common meds used in Pregnancy
- Magnesium Sulfate (acts at neuromuscular junction to slow impulses)
- Labetalol (Selective beta blocking agent)
- Hydralazine (Relaxes arterioles and ? vasospasm)
- Brethine / Terbutaline

Cardiopulmonary Hemodynamics

Swan-Ganz Pulmonary Artery Catheter
- Originally created in the 1970's to monitor LV function in patients undergoing surgery.
- Provides right- sided and left- sided heart pressures.
- Mixed venous (SVO2) blood sampling provides important data regarding oxygen delivery (DO2) and oxygen consumption (VO2).
- Provides central venous access and ease of blood sampling.

Indications for Insertion
- Hemodynamically unstable sepsis
- Complex, acute heart disease
- Acute, severe pulmonary disease
- Shock of all types if severe/prolonged
- Severe, multisystem trauma or large-area burn injury
- Major systems dysfunction

Contraindications for Insertion
- Severe coagulation defect or thrombolytic therapy
- Prosthetic right heart valve
- Endocardial Pacemaker

- Severe Vascular Disease
- Pulmonary HTN
- Significant Immune System Deficiency

Quadruple-Lumen PAC
Description of Catheter Lumens

Distal (PA) port-terminates at the catheter tip and measures PA and PA wedge pressures. Mixed venous blood samples may be drawn from this port.
- Balloon Inflation port-terminates at the tip of catheter, fills with 1.5 cc's of air.
- Proximal (RA) port-terminates within the right atrium, may be used to monitor CVP, admin fluids/meds and injection of solution for CO.
- Thermistor port- a temperature-sensitive wire that terminates near the tip of catheter.
- RV port-terminates in Rt. Ventricle, can be used for infusions and for pacing, not all have this.

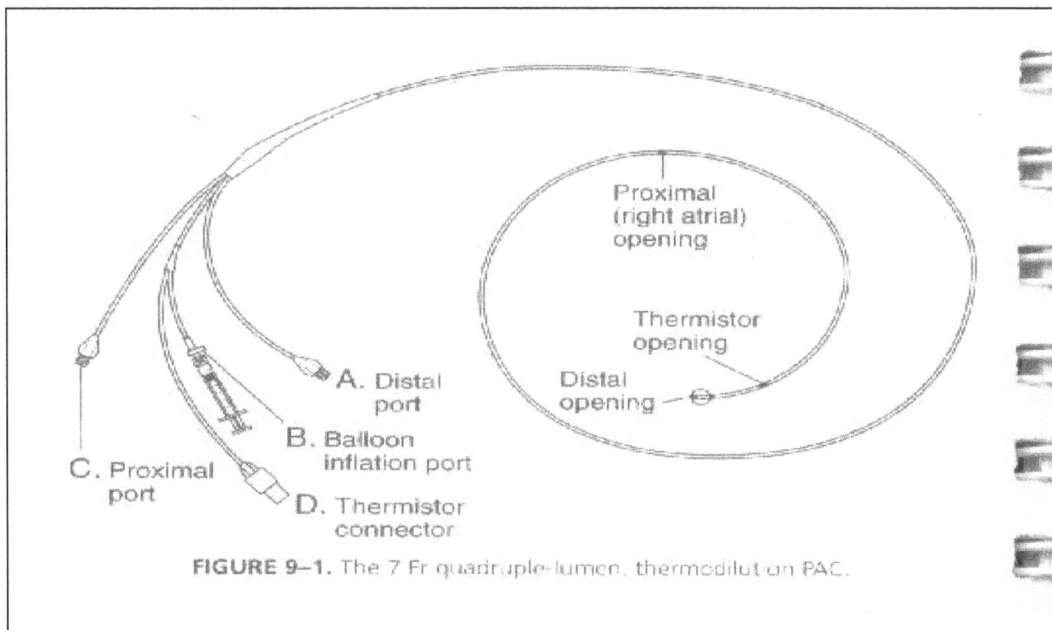

FIGURE 9-1. The 7 Fr quadruple-lumen, thermodilution PAC.

Waveform Tracings

- Passing of the pulmonary catheter from the right atrium, right ventricle, and terminating in the pulmonary artery

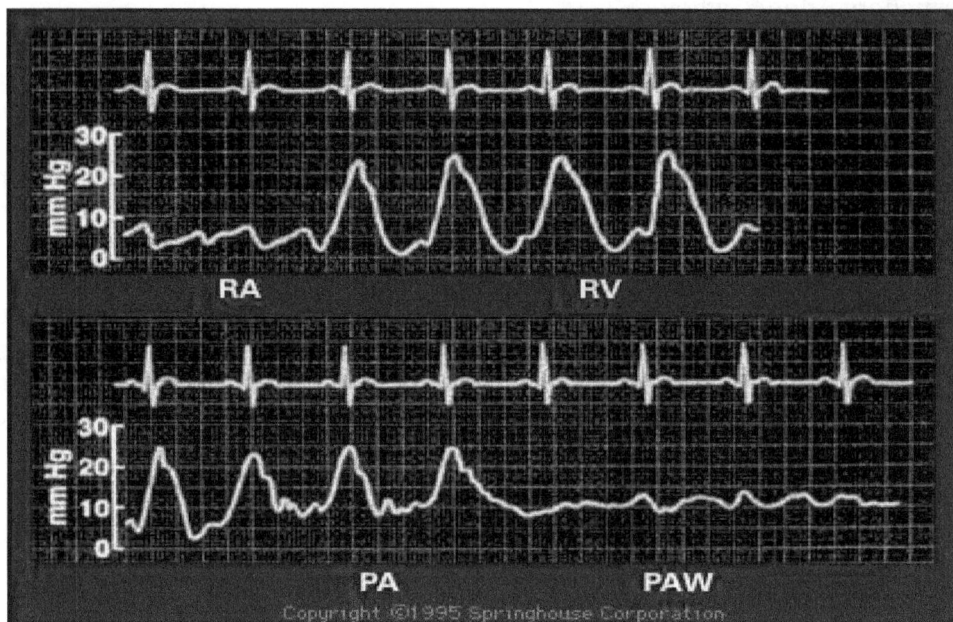

Right Atrial Waveform

RA Waveform / CVP Tracing

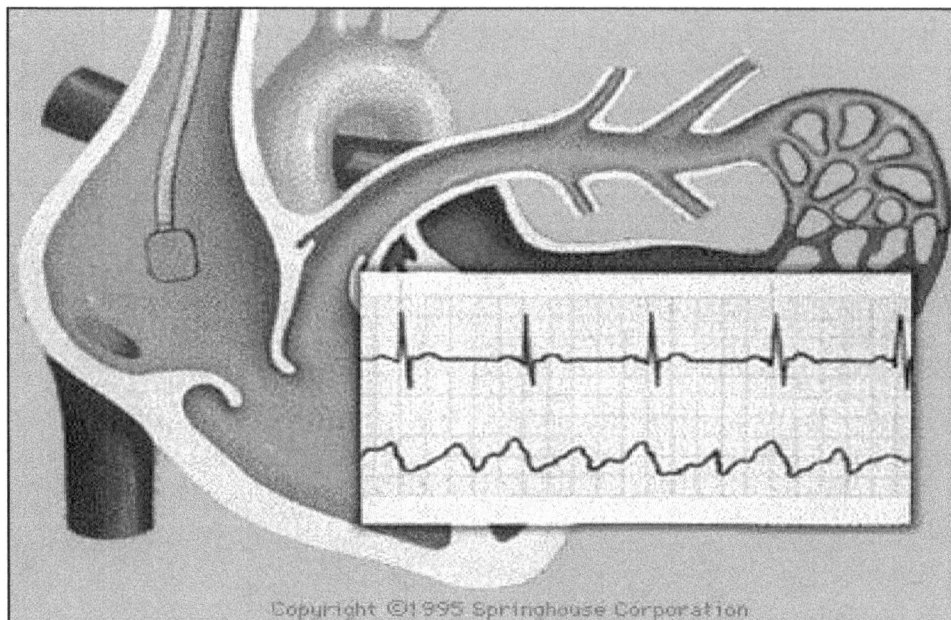

Right Ventricular Waveform

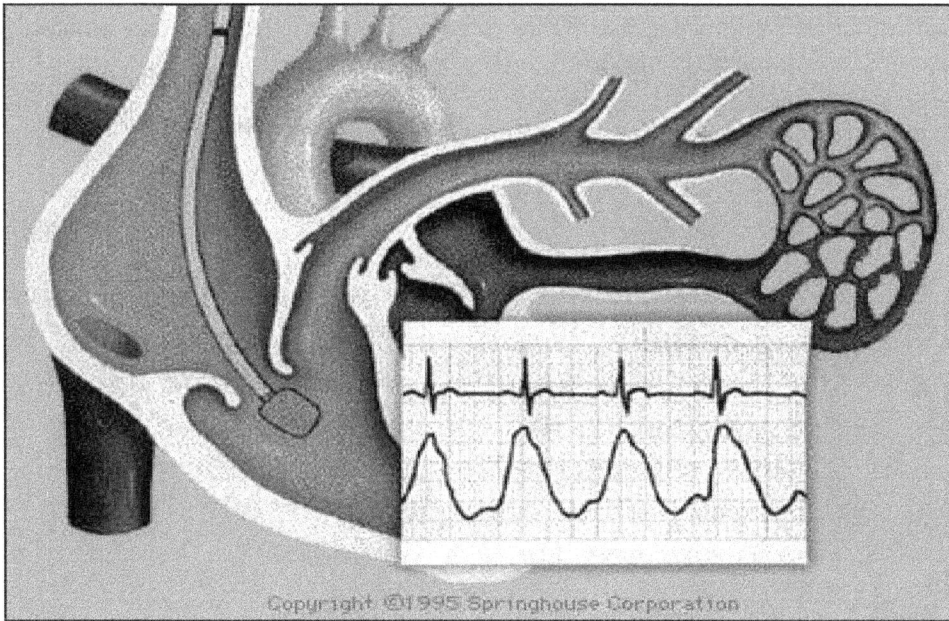

PA Waveform

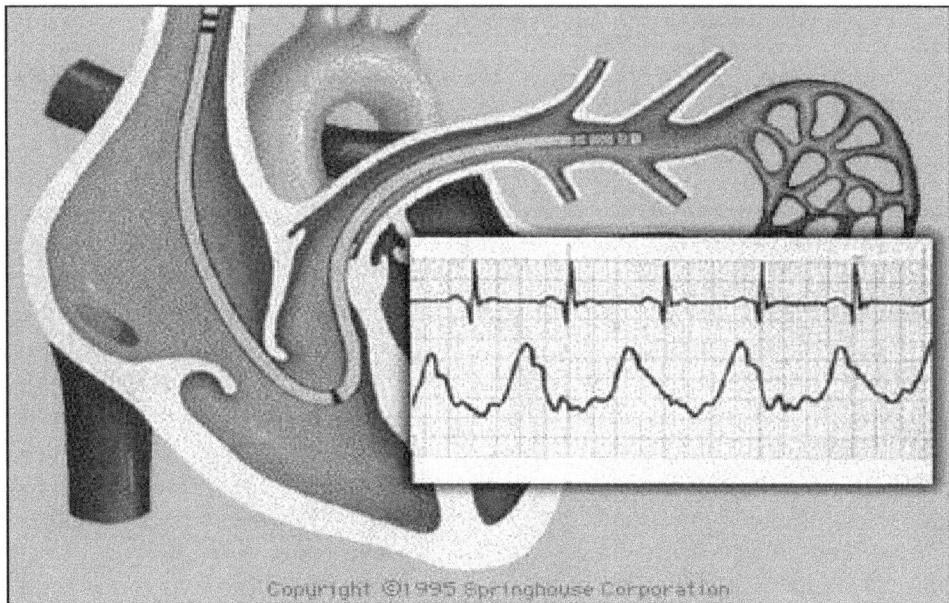

PCWP Waveform

Wedge Tracing
- If Patient is showing a RV Waveform, and the patients EKG is VT, the balloon needs to be "wedged" to float back into the PA. Even if the scenario shows the patient as "unstable". ie: Blood pressure 80/30, shortness of breath, etc.
- Monitor pt in V1 lead for detection of RBBB.
- Catheter-related dysrhythmias normally terminate with catheter withdrawal.

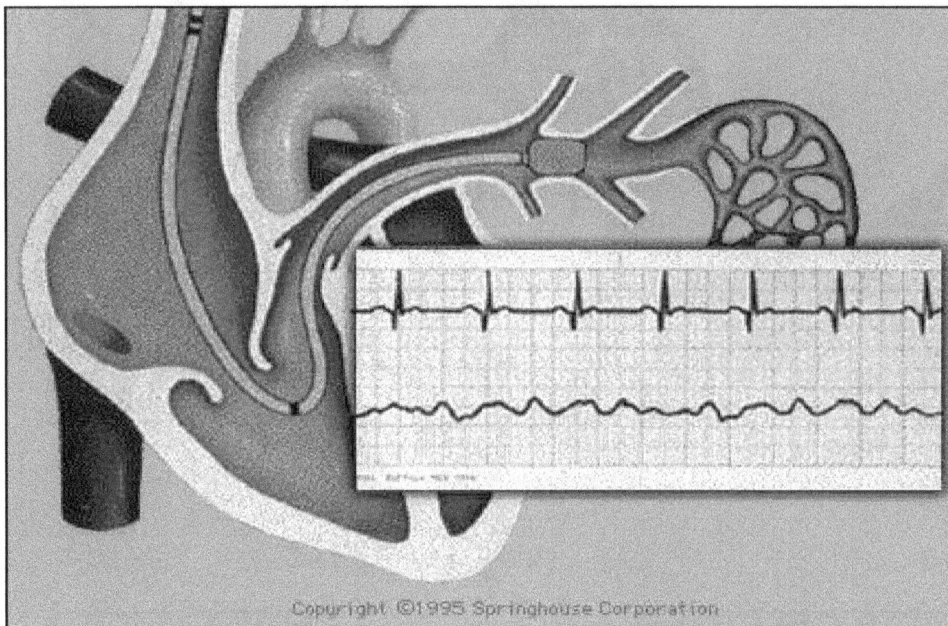

Copyright ©1995 Springhouse Corporation

Factors Influencing Systolic/Diastolic Pressures

Cardiopulmonary Pressures

1. CVP norm is 1-6mmHg
2. Cardiac index norm is 2.4 – 4.0 L/min/m2
3. PCWP norm is 6-12 mm/Hg

Control of Blood Pressure
- Arterial Baroreceptors, or pressure receptors, respond immediately to HTN by inducing reflex dilation of systemic arteries/veins, allowing the HR to slow. The opposite is true of hypotension.
- Chemoreceptor's immediately respond to hypoxemia, hypercarbia, and acidemia by inducing reflex systemic vasoconstriction as well as an increase in HR.
- Strong emotional stimuli, such as fear, pain, or anxiety result in tachycardia.

- Stroke Volume, or the amt of blood ejected /c each contraction, approx. 70ml, can increase/decrease in times of increased/decreased CO, HR, and contractility. Abnormal CO is most commonly r/t a problem /c SV.
- Any change in SV will normally produce a change in HR reciprocally, the exception to this is in pts /c an increase in metabolic rate, in which both SV/HR increase. The two most common reasons for a low SV is hypovolemia and LV dysfunction.
- Heart Rate. Generally, bradycardia/tachycardia is significant b/c they may reflect a potentially dangerous interference in CO.

Hemodynamic Pearls

1. Interpretation of normal pressures does not necessarily indicate normal cardiac functioning.
2. BP will not reflect early clinical changes in hemodynamic status.
3. SV will not typically fall until blood volume becomes too low, or the LV becomes too weak.

Factors Influencing **Stroke Volume**
1. Preload
2. Afterload
3. Contractility

Preload
- Determined by amt of blood remaining in the Lt. Ventricle at the end of diastole.
- Increases with greater venous return to the heart.
- Determines how much cardiac fibers will stretch.

Afterload
- The amt of pressure the Lt. Ventricle must work against to pump blood into the body.
- The greater the resistance, the more the heart works to pump out blood.
- Influenced by blood viscosity and resistance from valvular disease.
- Excessive afterload will increase LV work, decrease SV, increase myocardial O2 demands, and may result in LV failure.

Contractility
- The ability of muscle cells to contract after depolarization. This depends on how much the muscle fibers are stretched at the end of diastole.

Improving Cardiac Strength through Preload Reduction
- Preload reduction is done through vasodilation or with diuretics. Diuretics are commonly used for preload reduction mainly b/c they reduce excessive fluid in the circulatory system resulting from renal compensation for decreased blood flow through the kidney, for example, release of aldosterone, renin, and ADH.

- They work by blocking the reabsorption of Na and H2O, producing a rapid increase in UO.
- Vasodilators, such as NTG, diltiazem, and morphine can reduce preload as well by decrease the amt of blood returning to the heart.

Improving Cardiac Strength through Afterload Reduction
- Common in two situations: during hypertensive episodes and when the CO is low and the SVR is high.

Improving Cardiac Strength through Increased Contractility
- Inotropic therapy increases the strength of the cardiac contraction. This results
- In an increased EF, SV, CO, and ideally, tissue oxygenation.

Pharmacological Intervention

Common Inotropic Drugs

Dobutamine, 2-10 mcg/kg/min, onset 1- 2 min
Dopamine, 2-10 mcg/kg/min, onset <5 min
Amrinone, 5-10 mcg/kg/min, onset 5-10 min
Milrinone, .375-.75 mcg/kg/min, onset 5-5 min

Common Preload Reducers

NTG, 10-150 mcg/min, onset 1- 2 min
Diltiazem, .25mg/kg bolus, gtt 10mg/hr, onset 2-5 min
Lasix, 40-80 mg bolus, onset 5 min
Morphine, 5 mg bolus q5 min

Common Afterload Reducers

Nipride, .5-10 mcg/kg/min, onset 30-60 sec
Hydralazine, 10-40 mg bolus, onset 10-20 min
Vasotec, 1.25 mg bolus, onset 15 min
Lopressor, 5mg bolus q5 min x3 doses, onset 5 min
Labetalol, .25mg/kg q10 min initially, to a total dose of 50-300 mg, onset 5 min

Disease States Altering Hemodynamics
- Congestive Heart Failure
- Pulmonary Edema
- Hypertensive Crisis
- Cardiac Tamponade

- ARDS
- Sepsis
- Valvular Disease
  1. Cardiac Transplant
  2. Cardiogenic Shock

CHF

- LV failure usually precedes RV pump failure.
- Increased CVP is a late sign of LVF.
- Systolic dysfunction is characterized by a decrease in muscle strength.
- Diastolic dysfunction is characterized by inability of the ventricles to relax.

Cardiac Transplant

- Transplanted hearts may require sympathetic stimulants, such as Isoprel and neo-synephrine until the ventricles adjust to the absence of autonomic innervation.

Valvular Disease

- Mitral Regurgitation and Aortic Regurgitation both increase Preload
- Mitral Stenosis and Aortic Stenosis both increase afterload

Cardiac Tamponade

- Ventricular filling pressures increase and equalize as the heart is squeezed

ARDS/Sepsis

- Increased capillary permeability results in massive fluid depletion and a fluid-volume deficit, leading to hypotension.
- In ARDS, PCWP is usually normal.
- In early sepsis (warm shock), CO is normal to high and the SVR is decreased
- In later shock (cold shock); CO falls, while SVR may increase to off-set low BP.

Clinical Pearls

- Right-sided pressures should reflect the low-pressure vasculature of the lungs, while Left-sided pressures should reflect the higher pressures of the body.
- PAD/PCWP should be similar in absence of disease.

- PCWP is always lower than the PAD, except in pts with acute mitral insufficiency.
- Normally, for every 5cm H2O of PEEP, PWP increases by about 1mm Hg.
- PCWP is recorded at end-expiration.
- Large amts of PEEP may potentially increase intra-thoracic pressure, leading to decreased CO/Hypotension.
- Lung water begins to accumulate at wedge pressures of approx. 18-20 mm Hg.

Always Trouble-Shoot equipment first with undesirable readings!!!!

Check fluid levels, transducer level, adjust stopcocks, air bubbles in line, etc..

Safety with Swans

- Do not "wedge" the pt for longer than 15 seconds at a time, or 3 respiratory cycles!
- Withdraw air from balloon when done with CO!
- Do not "over wedge" pt when doing CO.
- Never infuse fluids/meds through PA distal port!
- Monitor PA tracing for inappropriate waveforms!
- Remember when infusing vasoactive gtts through proximal port to slowly flush or withdraw fluid from atrium before initiating a CO reading!

Transducer should be positioned where?

Should be zeroed at ground, level flight at altitude, and again upon landing.

Intra-Aortic Balloon Pump

Goals of IABP therapy include
1. Decrease the workload of the Heart
2. Decrease Myocardial Oxygen Demand
3. Increase Coronary Perfusion
4. Improve cardiac Output
5. Limits Myocardial Ischemia
6. Prevents Cardiogenic Shock

IABP utilizes
- Support in acute MI with Cardiogenic Shock
- Circulatory support in post-op CABG patients
- Intractable chest pain refractory to conventional care

IABP

- Inflation at the onset of diastole
- Deflation occurs just prior to the onset of systole

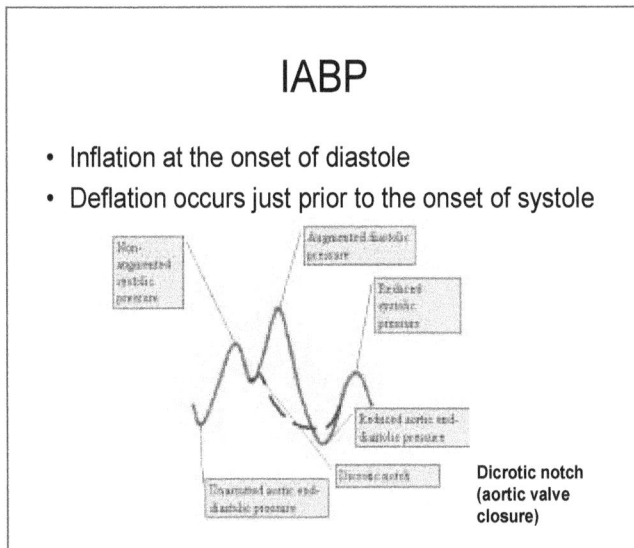

Dicrotic notch (aortic valve closure)

- Support in high-risk catherizations patients
- In severe ischemia as a bridge to revascularization

IABP Therapy
- Balloon inserted percutaneously in the femoral artery
- Balloon sits in the descending aorta, just distal to the left subclavian artery and above the renal arteries.
- Balloon inflates and deflates based on the patient's EKG or arterial pressure waveforms.
- Ventricular systole; the balloon is deflated
- Ventricular diastole; the balloon is inflated

Contraindications
- Aortic insufficiency
- Severe aortic disease
- Severe peripheral vascular disease

Complications
- Ischemia of limb distal to insertion site
- Aortic dissection
- Thrombocytopenia
- Septicemia
- Infection
- Renal complications
- Air / gas embolus

Early Inflation
- Inflation before the aortic valve closure
- Causes reflux of blood into the left ventricle
- Decreases Cardiac Output and increases SVR
- HARMFUL

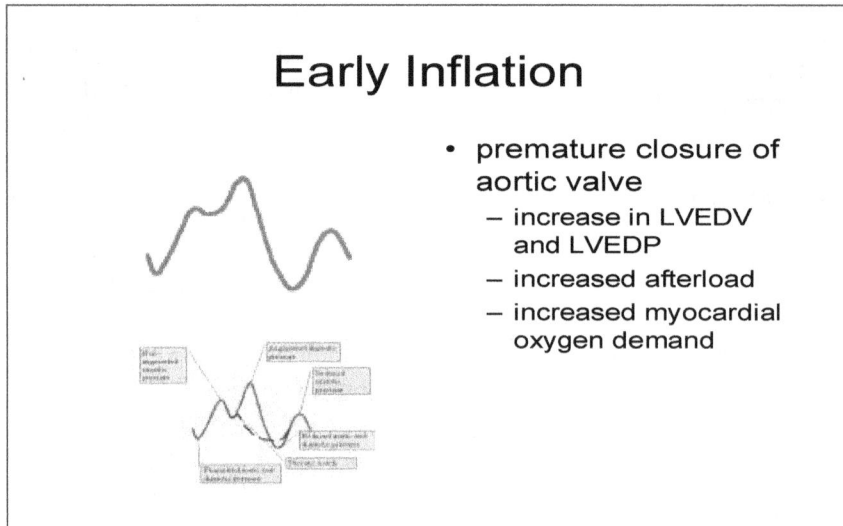

Late Inflation
- Results in Suboptimal Augmentation because of the minimal displacement of blood back towards the coronary arteries.

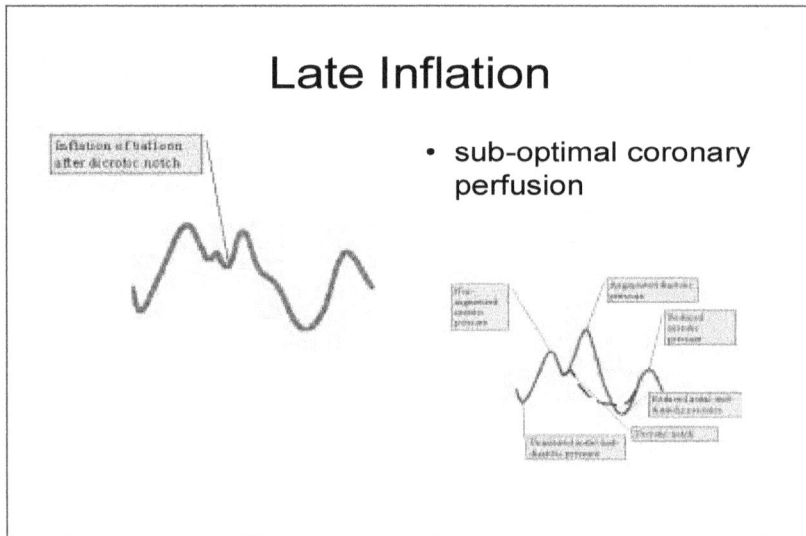

Early Deflation
- Vacuum effect and afterload reduction is lost.
- Occurs because by the time the aortic valve opens, the pressures in the aorta have equalized.

Late Deflation
- The Balloon is inflated during the beginning of ventricular systole.
- Increases the workload of the left ventricle.
- VERY HARMFUL to the patient!

Transport Considerations

- Due to the aircrafts vibration, you may need to switch your triggering mechanism to the arterial pressure line. Also, there may be too much artifact on the EKG to effectively trigger the balloon inflation / deflation.
- Due to Boyles Gas Law, you may find your balloon purging on ascent. This is due to the expansion of gases with increasing altitude.
- Balloon will purge again on descent due to gas contraction with decreasing altitude.
- Watch for patient decompensation during balloon purges. Be prepared to treat your patient.
- Make sure that all air is out of the art line to minimize dampening of the waveform since this may be a trigger source in the event of EKG trigger failure.
- In the event of Cardiac Arrest, place the trigger mode on arterial pressure or on internal trigger mode.
- In the event of power failure, the balloon needs to be manually pumped every 30 minutes to prevent thrombus formation on the balloon. TAKE THE SYRINGE!
- Ensure you have enough spare helium tanks. With the increasing changes in altitude and balloon purging, they may be necessary.
- Before leaving a facility, assure balloon placement by verifying the distal tip of the balloon is 2 cm. below the aortic arch, and the proximal end of the balloon does not occlude the renal arteries. Make sure to get a RECENT chest x-ray.
- Watch urinary output and distal pulses closely in transport.

Case Study Exercise

You are transporting a 55 year-old man with Cardiogenic Shock due to recent myocardial infarction. He is receiving intra-aortic balloon counterpulsation (IABP) in order to maintain adequate perfusion and hemodynamics.

You notice your patients mean arterial BP (MAP) has decreased from 70 to 45 mm Hg, and you suspect IABP malfunction. We know that we need at least a MAP of 60 to perfuse our kidneys, ensuring good renal flow. (MAP of 55 for cerebral perfusion) The action of the pump is timed from the patient's arterial BP tracing, and the balloon inflates with every beat. To diagnose this potential problem, you change the timing of balloon inflation to every third beat.

This is the patient's arterial pressure waveform... (Following page)

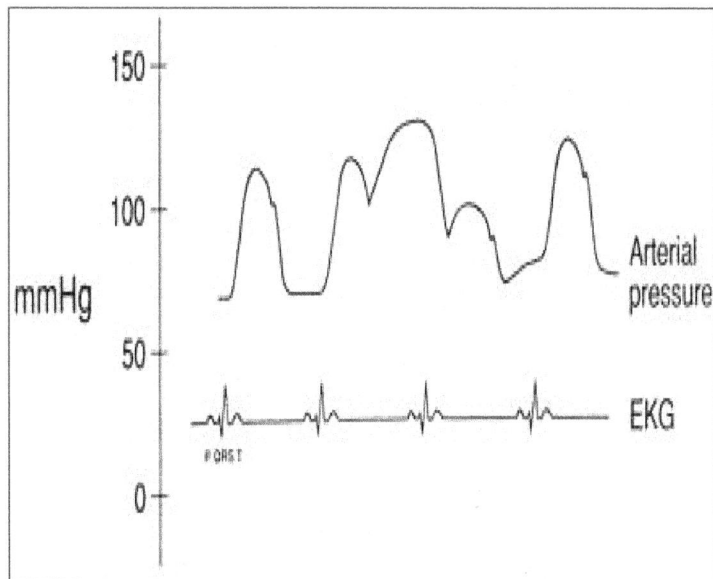

Which of the following is the most likely cause of the patient's decrease in BP?

    A. Early balloon inflation
    B. Early balloon deflation
    C. Late balloon inflation
    D. Late balloon deflation
    E. Problem is unrelated to balloon timing

Answer: **D Late balloon deflation**

The purpose of IABP is to inflate the balloon at the onset of diastole to augment diastolic pressure and coronary flow and to deflate the balloon prior to systole, thus reducing ventricular afterload. Optimal balloon inflation occurs at the dicrotic notch of the arterial BP waveform, and deflation should occur prior to the onset of ventricular systole. Deflation should occur at the onset of the ECG QRS complex, since the QRS is a marker of the electrical onset of systole.

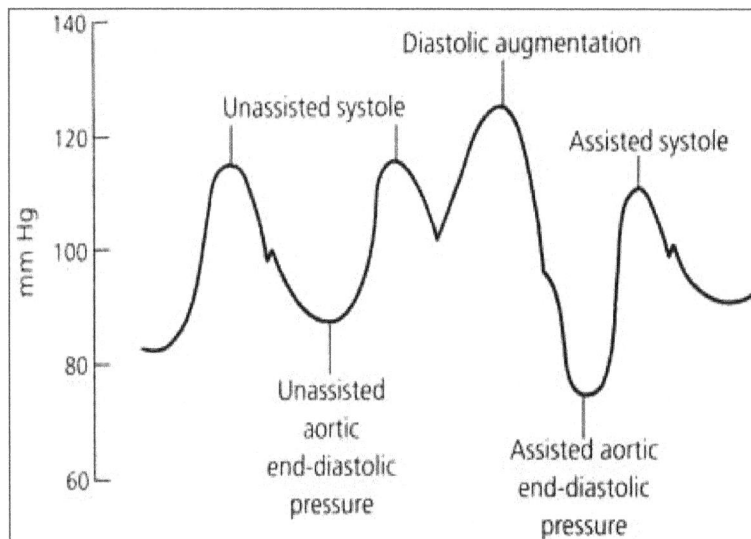

The arterial tracing from this patient demonstrates that balloon deflation is occurring late, after the onset of ventricular systole for the next beat as judged by the onset of the ECG QRS complex. This late deflation markedly increases both left ventricular afterload and myocardial oxygen demand, as the ventricle works against a balloon-augmented increase in end-diastolic pressure. This results in a decrease in systolic pressure in the beat following balloon inflation. The first beat and the last beat in the tracing, which do not follow balloon inflation, are not reduced. Properly timed balloon inflation and deflation are shown…

Review Questions

1. Your IABP begins to purge during ascent. The triggering mechanism for this function was initiated as a result of;
    a.   Boyle's Law
    b.   Gay-Lussac's Law
    c.   Charles's Law
    d.   Henry's Law

2. A patient was scuba diving and descended 66 feet. How many atmospheres of water pressure were on your patient?
    a.   1
    b.   2
    c.   3
    d.   None of the above

3. The percentage of oxygen at 25,000 MSL is;
    a.   4%
    b.   21%
    c.   18%
    d.   7%

4. The most common complication of fibrinolytic therapy is:
    a.   Dysrhythmias
    b.   Bleeding
    c.   Thrombosis
    d.   Hypocalcaemia

5. T-Wave inversion on an ECG usually is indicative of:
    a.   Infarction
    b.   Dysrhythmias
    c.   Injury
    d.   Ischemia

Acid Base Balance Interpretation

Control of Acid Base balance is essential for optimum functioning of chemical reactions, enzymes, and tissue oxygenation—i.e. LIFE.

Acid- Substance that donates a hydrogen ion

Base-Substance that accepts a hydrogen ion.

pH- Negative logarithm of the hydrogen ion concentration in which the degree of acidity or alkalinity of a substance is displayed.

Base Excess- Refers to the excess bases found in the blood. Metabolic Processes influence base excess.

Bicarbonate-Calculated value that is influenced by metabolic processes, and regulated by the kidneys.

Acid Base Disturbances

Acidemia- pH of the blood drops below 7.35. Results from too much acid or loss of base which upsets the ratio of 1:20 acid to base. Acidemia has a depressant effect of the CNS.

Alkalemia- pH of the blood greater than 7.45. Results from too much base or loss of acid, which again would upset the 1:20 ratio of acid to base. Alkalemia has a stimulant effect on the CNS.

Regulation of acid base imbalances

Buffers-
- Carbonic Acid System
- Protein Buffers
- Phosphate Buffers
- Respiratory System-
- Renal System (non- respiratory)

Correction vs. Compensation

Correction- All acid base parameters are restored to normal

Compensation-System that is causing imbalance continues to be affected, opposite

system begins to attempt to return pH to normal.

Acid Base Disorders

- Respiratory Acidosis
- Respiratory Alkalosis
- Metabolic Acidosis
- Metabolic Alkalosis

Respiratory Acidosis

ABG Values
- pH less than 7.40
- CO_2 greater than 40
- HCO_3 either normal or elevated

Causes
- Depression of respiratory center
- Respiratory muscle paralysis
- Chest wall injuries
- Disorders of the lung parenchyma

Signs and symptoms
- CNS depression
- Muscle twitching
- Arrhythmias, tachycardia, and diaphoresis
- Palpitations
- Flushed skin
- Electrolytes abnormalities

Respiratory Acidosis

Compensation

Kidneys increase hydrogen secretion and increase HCO_3 reabsorption.

Treatment
- Physical stimulation
- Vigorous pulmonary toilet
- Intubation & mechanical ventilation
- Reversal of sedatives/ narcotics
- Treatment of infection
- Diuretics for fluid overload

Respiratory Alkalosis

ABG Value
- Ph of greater than 7.40
- PCO_2 less than 40
- HCO_3 may be normal or decreased

- Psychogenic
- CNS stimulation
- Hypermetabolic states
- Hypoxia

Signs and Symptoms
- Headache
- Vertigo
- Parasthesias
- Tinnitus
- Electrolyte Abnormalities

Compensation
- Less hydrogen secreted
- More HCO3 is excreted

Treatment
- Treat pain /anxiety
- Correct hypoxia
- Treat fever
- Treat hyperthyroidism

Metabolic Acidosis

ABG Values:
- pH of less than 7.40
- PCO2 can be normal or decreased
- HCO3 will be less than 22 mEq/l.

Causes:
- Overproduction of acids
- Impaired excretion of acid
- Abnormal loss of HCO3
- Ingestion of acid

Signs and Symptoms
- CNS depression
- Cardiac Arrhythmias
- Electrolyte abnormalities
- Flushed skin
- nausea

Anion Gap
- Can be normal or elevated

Compensation:
- Ventilatory effort increases as acids accumulate

Treatment:
- DKA- fluids, electrolytes, and insulin
- Renal Failure-dialysis

- Low cardiac output-increase tissue perfusion

Metabolic Alkalosis

ABG Values:
- pH greater than 7.40
- PCO2 can be normal or increased
- HCO3 greater than 26 mEq/l

Causes:
- large loss of gastric contents
- Loss of potassium
- Ingestion of excess bicarbonate
- Prolonged use of diuretics

Signs and Symptoms:
- Diaphoresis
- Nausea/ vomiting
- Increased neuromuscular activity
- Shallow respirations
- EKG changes
- Confusion to coma
- Electrolyte abnormalities

Compensation:
- Decrease in respiratory rate to retain CO2.
- Kidneys will increase the HCO3 lost in the urine.

Treatment:

- Replace lost potassium, use 0.9 % sodium chloride
- Diamox

Normal pH	7.35-7.45
Normal PCO2	35-45
Normal HCO3	22-26
Compensation	pH to normal
Uncompensated	pH not normal
	Hco3 / PCO2 are normal
Partially Compensated	ALL values are abnormal
Combined	All values are either acid or base

ABG Interpretation

Step 1

$$\frac{pH + CO_2}{}$$
RESPIRATORY PROBLEM

Example

pH—7.30
PCO2—48
HCO3—26
= Respiratory Problem (Respiratory Acidosis)

Step 2

$$\frac{pH + HCO3}{}$$
METABOLIC PROBLEM

pH   7.51
PCO2 40
HCO3 33

METABOLIC PROBLEM (Metabolic Alkalosis)

Step 3

Compensated?
Partially Compensated?
Combined?

Sample A

pH   7.37
PCO2 53
HCO3 32
Compensated Respiratory Acidosis

Sample B

pH   7.29
PCO2 37
HCO3 19
Uncompensated Metabolic Acidosis

Oxyhemoglobin Dissociation Curve

The O2 Dissociation curve is really a love story…

<u>Left Shift = Lovers</u> The Hemoglobin molecules love oxygen and will not readily release it

<u>Right = Readily Releases</u> The Hemoglobin readily releases the oxygen molecules for use

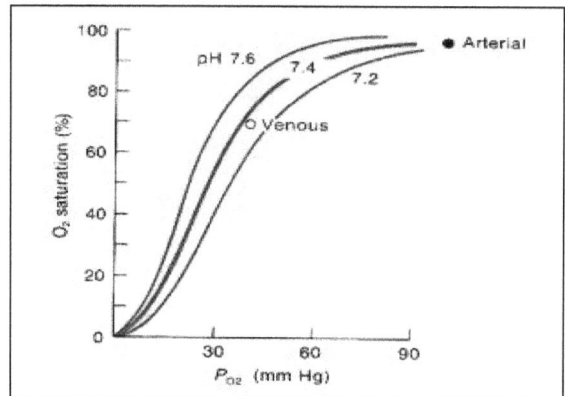

Practice Problems

Pt with a temp of 102:	L	R
Pt with a CO2 of 30:	L	R
Pt with ↓ 2, 3, DPG:	L	R
Hemophiliac:	L	R
pH of 7.50:	L	R
CO2 of 50:	L	R
Frozen lake drowning:	L	R
Ph. of 7.20	L	R

Remember… Left = ⬇ Low

Low	Right
H+ = Alkalosis	Acidosis
CO2	Hypercarbia
Temp	Hyperthermia
2,3,DPG	2,3,DPG

Rules to live by:

- For every 10mm/Hg change in CO_2, the pH will change <u>0.08 in the opposite direction</u>

- For every 10mEq change of Bicarb, the pH will change <u>0.15 in the same direction</u>

Burn Management

Current Trends in Burn Treatment

Demographics

Burns are the third leading cause of death and injury in children

Initial Assessment and Management

Primary Survey

Airway and Smoke Inhalation
- 20% -50% of all admissions
- 60% -70% of all deaths
- 3 types of inhalation injury

Carbon monoxide poisoning
- Injury above the glottis
- Injury below the glottis

Carbon Monoxide Poisoning
- Most common cause of pre-hospital mortality
- 200X greater affinity for hemoglobin
- Shifts oxygen dissociation curve to the left

Signs and Symptoms
- PaO_2 is unaffected (oxygen dissolved in plasma not affected)
- Cyanosis and Tachypnea not present (CO_2 removal not affected)
- >60% death
- 40% -60% obtundation
- 15% -40% CNS dysfunction
- 5% -10% smokers

Carbon Monoxide Treatment
- Oxygen
- Room Air
- 100% FiO2
- HBO*

Half Life of CO
- 250 minutes
- 60 minutes
- <30 minutes

Injury above the Glottis

Injury below the Glottis

Choices For Access
- First choice: Peripheral vein; nonburn area
- Second choice: Peripheral vein; burn area
- Third choice: Central vein; nonburn area
- Worst choice: Central vein; burn area

Fluid Resuscitation

REMEMBER THIS FORMULA:

4ml/kg/BSI with the first half given in the first 8 hours, the remaining administered in the next 16 hours.

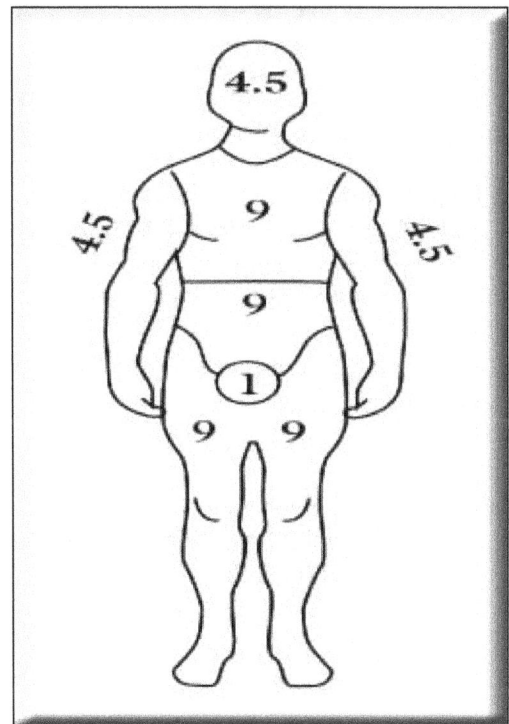

Pediatric Burns
- Greater body surface area to body weight ratio

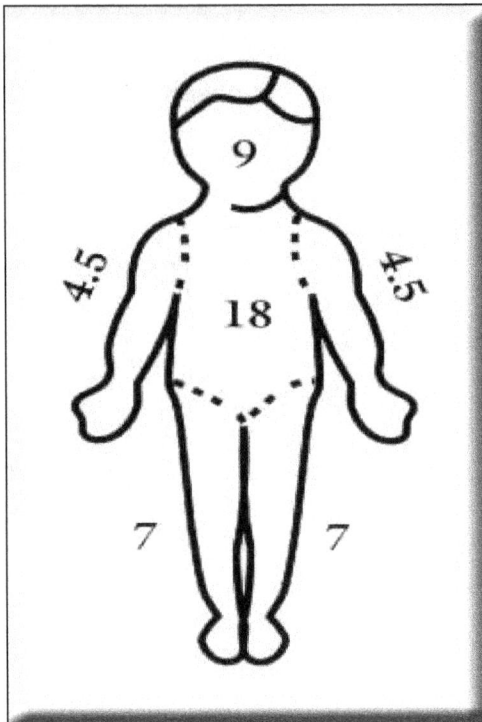

Fluid Resuscitation
- Use D5LR for children < 5 years old
- Use LR for children >5 years old
- 4 cc x TBSA x weight in Kg =

Total amount of fluid required in the First 24 hours post burn injury
Give ½ of this amount in the first 8 hours

Pediatric Parkland Formula
- 5 yr old --weight = 23 Kg with a 65% second and third degree
- 4cc x 23 kg x 65%TBSA=5980cc
- 5980 / 2 = 2990cc
- 2990 / 8 = 374cc/Hr x 8 Hrs

Increased Vascular Permeability
- Altered microcirculation from direct heat injury and inflammation
- Increased protein permeability leading to large plasma leak
- Accumulation of protein rich edema below eschar
- Hypovolemia

Vitamin C (Ascorbic Acid)
- 37 pts >30% TBSA hospitalized within 2 hours of injury

Results:
- resuscitation fluid volume requirements
- body weight gain
- burn wound edema
- severity of respiratory dysfunction

Mechanism of action
- Water soluble antioxidant
- Reduces post-burn lipid peroxidation
- Scavenges free radicals
- ? Osmotic diuretic effect

Dosage
- 66 mgs/kg/hr
- Start immediately

Physiologic Response to Burn Injury

Pathophysiology of Burn-Induced Acute Lung Injury
- Early, hypoproteinemia contributes to extravasations of fluid into the interstitium

Inflammatory:
- Complement, neutrophils and oxygen-derived free radicals
- Neutrophils sequestration within the microvasculature
- TNF-alfa release likely increases permeability

Cardiovascular
- Tachycardia
- Cardiac output
- Catecholamine storm
- Hypotension
- Does not always=under resuscitation
- Low dose Vasopressors
- Monitor PCWP, CVP, CO

Metabolic Alterations
- Early electrolyte abnormalities from skin losses and fluid shifts (.Ph, .Mg, .Ca, .K+)
- Acute micronutrient losses
- Selenium, Vit C, chromium
- Fever

- 1st 72o due to cytokine production

Metabolic Alterations
- Acidosis
- Usually reflects under resuscitation
- Other possibilities
- CN/CO poisoning
- Massive tissue injury
- Comorbid cardiac dz
- Tissue edema (compartment syndrome)

Hematologic Alterations
- Hemoconcentration
- Thrombocytopenia
- Leukocytosis with left shift (inflammatory mediators)
- Elevated PT

Renal Alterations
- Rhabdomyolysis
- Tissue (muscle) injury
- Compartment syndrome
- Deep gasoline burns
- Poor perfusion-predisposes to ATN
- Large urine output not absolute requirement-may lead to tissue edema

Gastrointestinal Alterations
- Early gastroparesis/food retention
- Stress related mucosal disease (Curling's Ulcers)
- Dysmotility-stomach/colon
- Acute malnutrition

Endocrine Changes
- Hyperglycemia (insulin resistance)
- Parathyroid reduction (hypocalcemia)
- Thyroid abnormalities (low free T3)
- Adrenal insufficiency

Immune Alterations
- Acute immunodeficient state
- High risk of infection
- Tissue injury leads to nidus for infection
- Inhalation injury/mechanical ventilation predisposes to VAP
- Multiple device sources
- Predisposed to multi-resistant organisms and fungal invasion due to prolonged hospital stay-especially with large burns

Surgical Management
- Burn Injury
- Assessment of the Burn Wound

IT ONLY TAKES SECONDS

Temperature of Water vs. Time to Burn?
- 150 Degrees   2 Seconds
- 140 Degrees   6 Seconds
- 125 Degrees   2 Minutes
- 120 Degrees   10 Minutes

Categories of Burns
- Thermal – 85%
- Electrical – 10%
- Chemical – 5%

Tissue Injury is Dependent on:
- Voltage of the source
- Amperage of current passing through the tissues
- Resistance of tissue traversed by current
- Duration of contact
- Pathway of the current

Chemical Burns
- More than 30,000 different chemicals
- 5% of all burn admissions
- Severity of injury depends on duration of contact
- Concentration of chemical

Chemical
- Common Agents
- Hydrochloric Acid-Muriatic Acid
- Formic Acid
- Chromic Acid
- Hydrofluoric acid
- Sodium Hydroxide
- Potassium Hydroxide

Emergency Care
- Remove all Clothing
- Brush off the skin if the agent is a powder
- Irrigate
- Strong chemicals can contaminate large quantities of water (10ml of 98% sulfuric

acid will reduce the pH of 12 liters of water to a pH of 5)
- NO NOT try to Neutralize

Treatment of the Burn Wound

Topical Therapy
- Silver Sulfadiazine
- Mafenide Acetate
- Silver Nitrate

Skin Substitutes
- Bilaminates
- Xenografts
- Allografts
- Transcyte
- Integra

Cultured Epidermal

Autograft
- Requires biopsy
- 3 to 4 weeks to delivery
- Expensive: $1,000 for 50 sq cm
- Prolonged Immobilization
- Thin and Friable
- Large full thickness burns

Conclusion
- Burn injury results in physiologic changes of many organ systems, not just the skin
- Effective burn resuscitation remains a challenge
- Airway injury and smoke inhalation greatly increase mortality and initial effective management is crucial
- Electrical and chemical injuries have special considerations
- Management of the burn wound in a burn center has many benefits including reduction in length of hospital stay, earlier return to work, and optimal functional and cosmetic results

Review Questions

1.   Your patient has 2nd degree burns to the anterior chest, left arm, and both legs. He has been in the local ER for approximately 6 hours with a total infusion of Lactated Ringers of 500ml. How much fluid will you need to infuse during your 2 hour fixed wing flight to Alberta, Canada?

Neurological Emergencies

Outline / Objectives
- Review Anatomy and Physiology of the Nervous System
- Review the components of the Neurological Exam
- Review the physical exam
- Review documentation of findings
- Review findings of a few abnormal conditions and disease states

Anatomy of the Neurological System
- Central Nervous System (CNS)
- Brain and Spinal Cord
- Peripheral Nervous System (PNS)
- Motor & sensory nerves outside the CNS
- Autonomic Nervous System
- Regulates internal environment – involuntary control
- Sympathetic nervous system
- Parasympathetic nervous system

Brain
- Two internal carotid arteries
- Two vertebral arteries
- Basilar artery
- Three major units
- Cerebrum
- Cerebellum
- Brainstem

Cerebrum
- Cerebral cortex – grey outer layer
- Frontal lobe – vol. skeletal / fine motor
- Parietal lobe – sensory / awareness of body position (proprioception)
- Occipital lobe – visual center
- Temporal lobe – sound, taste, smell, balance
- Limbic system - behavior

Cerebellum
- Coordination of voluntary movement
- Utilizes sensory data for reflexive control of muscle tones, equilibrium, and posture

Brain Stem
- Medulla oblongata
- Pons
- Mid-brain
- Diencephalon
- Nuclei of the 12 cranial nerves
- Thalamus

Structures of the Midbrain

Diencephalon
- CN I – II
- Thalamus
- Relays impulses between parts of the brain
- Coveys all sensory impulses (except olfaction) to and from the cerebrum
- Controls consciousness, sensations, abstract feelings

Midbrain
- CN III – IV
- Reflex center for eye and head movement
- Auditory relay pathway
- Corticospinal tract pathway

Pons
- CN V – VIII
- Reflexes of pupillary action and eye movement
- Regulates respiration and houses a portion of the respiratory center
- Controls voluntary muscle action

Medulla oblongata
- CN IX – XII
- Respiratory, circulatory, and vasomotor activities
- Houses the respiratory center
- Reflexes for swallowing, coughing, vomiting, sneezing, and hiccupping
- Relay center for ascending and descending spinal tracts

Epithalamus
- Pineal body
- Sexual development and behavior

Brain Stem
- Nuclei of the 12 cranial nerves

Hypothalamus
- Major center for internal stimuli for autonomic nervous system
- Temperature control, water metabolism, body fluid osmolarity, feeding behavior, and neuroendoricine activity

Pituitary
- Hormonal control of growth, lactation, vasoconstriction, and metabolism

Cranial Nerves				
I	Olfactory	Sensory	Oh	Some
II	Optic	Sensory	Once	Say
III	Occulomotor	Motor	One	Marry
IV	Troclear	Motor	Takes	Money
V	Trigeminal	Both	The	But
VI	Abducens	Motor	Anatomy	My
VII	Facial	Both	Final	Brother
VIII	Vestibulocochlear	Sensory	Very	Says
IX	Glossopharyngeal	Both	Good	Bad
X	Vagus	Both	Vacations	Boys
XI	Spinal Accessory	Motor	Seem	Marry
XII	Hypoglossal	Motor	Heavenly	Money

The Cranial Nerves

CN I Olfactory
Sensory: smell reception and interpretation

CN II Optic
Sensory: visual acuity and visual fields

CN III Oculomotor
Motor: raise eyelids, most extra ocular movements

Parasympathetic: pupillary constriction, change lens shape

CN IV Troclear
Motor: downward, inward eye movement

CN V Trigeminal
Motor: jaw opening and clenching, chewing and mastication

Sensory: sensation to cornea, iris, lacrimalgland, conjunctiva, eyelids, forehead, nose, nasal and mouth mucosa, teeth, tongue, facial skin

CN VI Abducens
Motor: lateral eye movement

CN VII Facial
Motor: facial expressions, close eyes
Sensory: taste (anterior 2/3 of tongue), pharynx
Parasympathetic: secretion of saliva and tears
CN VIII Acoustic
Sensory: hearing and equilibrium

CN IX Glossopharyngeal
Motor: muscles for swallowing and phonation
Sensory: nasopharynx, gag reflex, posterior 1/3 of tongue
Parasympathetic: secretion of salivary glands and carotid reflex

CN X Vagus
Motor: phonation and swallowing
Sensory: behind ear and external ear canal
Parasympathetic: secretion of digestive enzymes, peristalsis, carotid reflex, involuntary action of heart, lungs and digestive tract

CN XI Spinal Accessory
Motor: turn head, shrug shoulders, some phonation

CN XII Hypoglossal
Motor: tongue movement for speech and swallowing

Spinal Nerves
- 31 pairs
- Exit at each intervertebral foramen
  1. spinal cord
  2. dorsal root ganglion (one of several visible)
  3. rootlets of spinal nerves
  4. vertebral artery
  5. spinal nerve
  6. dura (reflected)

Dermatome
- Often some overlap

Infants and Children
- Major portion of brain growth in 1st year – brain continues to grow until age 12 - 15
- Myelinization of brain and nervous system – does not respond well to injury/disease
- Primitive reflexes in newborn:
- Yawn, sneeze, hiccup, blink at bright light and loud sound, pupillary constriction with light, withdraw from painful stimuli
- Motor maturation in cephlocaudal direction

Pregnancy
Physiologic alterations
- Contraction or tension headaches
- Numbness and tingling of the hands (postural kinking of blood vessels at thoracic outlet) – often confused with Carpal tunnel syndrome
- Sleep patterns disrupted in early and late pregnancy

Older Adults
- The Bad News

Physiologic alterations
- Cerebral neurons decrease by 1% per year beginning at age 50

The Good News
- The velocity of nerve impulse conduction declines 10% between age 30 and 90 – so, responses to various stimuli take longer – therefore Your Boss will not make any "Snap" decisions!

Review of Related History
- History of Present Illness (HPI)
- Seizures
- Sequence of events / aura / fall / motor activity / loss of consciousness / postictal phase
- Muscle tone
- Frequency / medications
- Pain
- PQRST / medications
- Gait
- Balance / falls / arthritis / medications

Review of Related History
- Past Medical History (PMH)
- Trauma / infections / congenital / CV / surg
- Family History
- Hereditary issues / alcoholism / mental retardation / epilepsy / Alzheimer's / learning disorders / thyroid disease / HTN
- Personal / Social History
- Environmental/occupational hazards, sleeping patterns,
- use of alcohol or recreational drugs

Physical Exam of Cranial Nerves

Cranial Nerves (typically II – XII)
Proprioception and Cerebellar Function

Sensory Function
- Deep Tendon Reflexes
- CN II (Optic) Test vision / peripheral visual fields
- CN III, IV, and VI (Oculomotor, Troclear, and Abducens)
- Inspect eyelids for drooping; pupils; extra ocular eye movements
- CN V (Trigeminal)

- Inspect face for atrophy; palpate jaw muscles (clenched teeth); sensory in each extremity
- CN VII (Facial) Symmetry of face with expressions (smile, frown, puffed cheeks, and wrinkled forehead) Sweet and salty taste on tongue (hard to do in pre- hospital)
- CN VII
- CN VIII (Acoustic) Sense of hearing
- CN IX (Glossopharyngeal) Test gag reflex and ability to swallow
- CN X (Vagus) Speech sounds (presence of nasal or hoarse quality to voice), swallowing difficulty
- CN XI (Spinal Accessory) Test trapezius muscle (shoulder shrug against resistance) Test sternocleidomastoid muscle strength (turn head side to side against resistance)
- CN XII (hypoglossal) Test tongue movement and strength up/down/side-to-side

Proprioception and Cerebellar Function
- Rapid rhythmic alternating movements
- Hands up/down
- Finger to finger / nose to finger
- Balance and gait

Sensory Function
- Patient's eyes closed
- Minimal stimulation – increase until patient feels stimulation
- Evaluate correct interpretation – sharp/dull/warm/cold/dry/wet

Patterns of Sensory Loss
- Single peripheral nerve
- Multiple peripheral nerves (polyneuropathy) – often extremity
- Multiple spinal nerve roots – incomplete loss of sensation
- Complete transverse lesion of the cord – all sensation lost below lesion
- Partial spinal sensory syndrome (Brown-Sequard Syndrome) – sensory

(pain/temp) loss on one side – proprioception/motor loss on lesion side

Reflexes (Superficial)
- Abdomen – with patient lying supine, stroke each quadrant of the abdomen
- A slight movement of the umbilicus toward each area of stimulation should be bilaterally equal
- Plantar reflex – stroke the lateral side of the foot, from the heel to the ball; then curve across the ball to the medial side
- The patient should have plantar (downward) flexion of all toes
- The Babinski sign is present when there is dorsiflexion (upward) of the great toe with or without fanning of the other toes
- This is expected in children < 2 yrs.
- Indicates a Neuro defect in adults

Deep Tendon Reflexes
- Focus patient's attention on another muscle group (i.e. pulling clenched hands apart)
- Position limb with slight tension on the tendon to be tapped
- Palpate the tendon to determine the correct area of stimulation instead of randomly tapping all areas
- Compare each side to the other – not one tendon to another
- Document findings as 0 to 4 + (0=none, 1=sluggish, 2=normal, 3=slightly brisk, 4=hyperactive)
- Test for ankle clonus
- Support the patient's knee in a partially flexed position
- Briskly dorsiflex (upward) the patient's foot, maintaining the foot in flexion
- No rhythmic oscillating movements should be palpated
- Sustained clonus is associated with upper motor neuron disease

Meningeal Signs
- Nuchal rigidity (stiff neck) is associated with meningeal inflammation and intracranial hemorrhage
- Brudzinski's Sign – involuntary flexion of the hips and knees when flexing the neck

- Kernig's Sign – patient in supine position, flex leg at knee and hip – pain in the lower back and resistance to straightening the leg at the knee constitute a positive sign

Posturing
- Common Abnormalities
- Skull Fractures

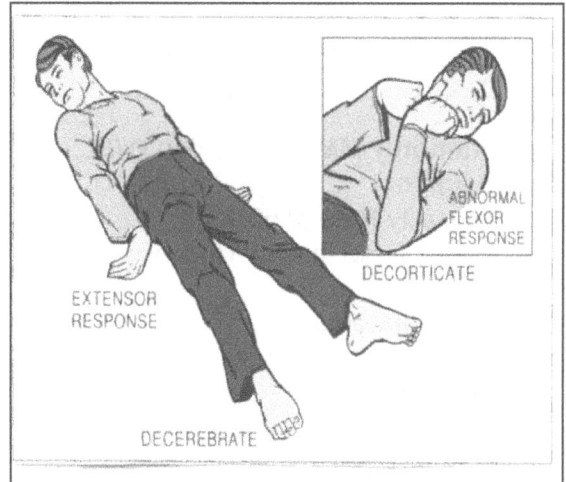

Location of brain involvement
- De-corti-cate = Cortex of the Brain
- De-Cere-Brate = Cerebellum

Signs of **Basilar Skull Fracture:**
- Raccoon's Eyes (Periorbital Ecchymosis)
- Battle's Sign (Retroauricular Ecchymosis)
- Cerebral Spinal Fluid leakage
- Halo test
- Glucose dipstick
- Basilar Skull Fracture
- Epidural Hematoma
- Outside dura/within skull

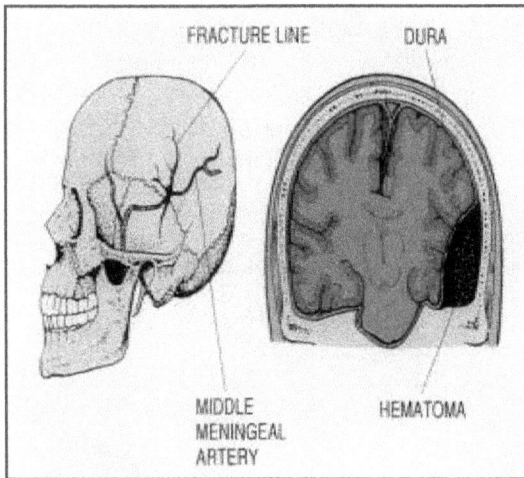

Most commonly MMA Tear *(You WILL see this one again!)*
- Mortality Rate of 50%
- +LOC/Lucid Interval/Unconsciousness

Subdural Hematoma
- Located under the dura but above the arachnoid space
- Typically a rupture of the veins in subdural space.
- Onset can be acute or in 24 – 48 hours post injury.

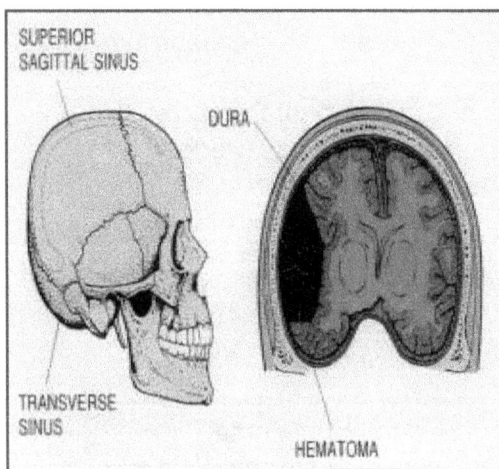

Subarachnoid Hemorrhage
- Medical
- Aneurysm in nature
- Higher incidence of seizures, vasospasm, rebleeding

Trauma
- Different etiologies
- Bleeding from cortical contusion through pia mater

- Disruption of vessels with shearing forces
- Bleeding initially through ventricles

Common Abnormalities
- Coma-is a profound state of unconsciousness.
- A comatose patient cannot be awakened, fails to respond normally to pain or light,
- Does not have sleep-wake cycles, and does not take voluntary actions.
- Brain trauma, the most common cause of comas, accounts for 60% of cases
- Various toxins can also lead to coma, including poisons, alcohol, barbiturates, opiate narcotics, sedatives, amphetamines, cocaine and aspirin.
- Metabolic abnormalities :diabetes ,liver or kidney failure, hypoxia and electrolyte imbalances

Brain Stem abnormalities
- Damage to Reticular Activating System (RAS)
- Structural lesions

Assessment of Comatose Patient

The goal of a neurological examination in a comatose patient is to determine if the coma is induced by
- a structural lesion
- metabolic derangement
- Possibly from both.
- Mental status is evaluated by observing the patient's response to visual, auditory and noxious (i.e., painful) stimuli

Pupillary Response
- Pupil size and reactivity might suggest a possible etiology or localize the brain lesion responsible for coma.
- Most metabolic encephalopathy's produce small but reactive pupils.
- Atropine, scopolamine, glutethimide, hypoxia, hypothermia and severe barbiturate intoxication produce large unreactive pupils.
- Midsize unreactive pupils suggest midbrain lesion.

- Pinpoint pupils suggest a pontine lesion or opiate overdose
- A small pupil accompanied by a Horner's syndrome suggests a hypothalamic lesion.
- A large dilated fixed pupil suggests IIIrd nerve compression from incipient herniation or aneurysm.

Breathing Patterns

- Cheyne-Stoke respiration consists of slow periods of hyperventilation that alternate with periods of hypoventilation. This respiratory pattern occurs in patients with bilateral cerebral disease or heart failure.
- Central Neurogenic hyperventilation is rapid breathing (40-70 respirations/minute) and occurs in patients with midbrain disease
- Apneustic breathing is slow breathing with a prolonged pause between inspiration and expiration.

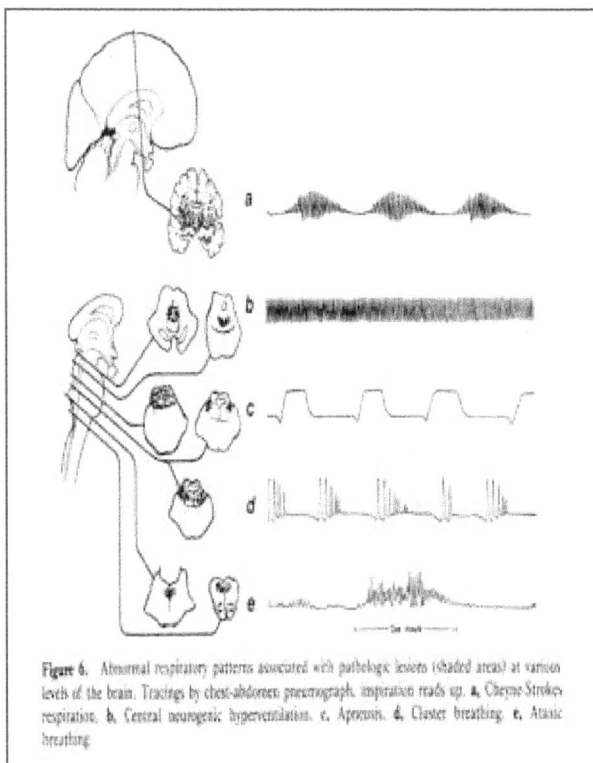

Figure 6. Abnormal respiratory patterns associated with pathologic lesions (shaded areas) at various levels of the brain. Tracings by chest-abdomen pneumograph, inspiration reads up. a. Cheyne-Stokes respiration. b. Central neurogenic hyperventilation. c. Apneusis. d. Cluster breathing. e. Ataxic breathing

Increased ICP

- headache,
- visual disturbances,
- nausea,

- vomiting,
- altered level of consciousness,

Cushing's triad

- The body's response to a rise in ICP.
- Involves an *increased systolic blood pressure*
- Increasing difference between systolic and diastolic blood pressures
- *Decrease in pulse rate*
- *An abnormal respiratory pattern*
- Irregular respirations occur when injury to parts of the brain interfere with the respiratory drive

Understanding the Monroe Kellie Hypothesis is the key to understanding and managing ICP

1. Cluster breathing occurs as clusters of breaths followed by apneic periods of variable duration.

2. Apneustic and cluster breathing occur in patients with pontine disease

3. Ataxic breathing consists of an irregular respiratory rate and rhythm.

4. Ataxic breathing occurs in patients with medullary disease.

Seizures

Partial seizures
- Characterized by onset in a limited area, or focus, of one cerebral hemisphere.

Complex partial seizures
- Temporal lobe often begins with a motionless stare followed by simple oral or motor automatisms.
- Frontal-lobe seizures often begin with vigorous motor automatisms or stereotyped clonic or tonic activity.

Treatment Modalities:
- ABC
- RX
- Valium
- Versed
- Ativan
- Phenytion
- Phenobarbital

Flight considerations
- Simple
- ABC
- Patient Safety
- Monitor and intervene when required

Common Abnormalities

Multiple Sclerosis
- Degenerative disorder – break-down of blood-brain barrier and allows immune cells to pass into myelinated white matter – myelin is destroyed and axons no longer transmit signals
- Fatigue, bowel/bladder dysfunction, sexual dysfunction, sensory changes, muscle weakness, cognitive and emotional changes
- Age 20 to 40
- Ratio women to men 2:1

Meningitis
- Inflammation of the meninges – bacteria / virus
- Fever, chills, nuchal rigidity, headache, seizures, and vomiting
- Young infants do not exhibit nuchal rigidity until 6 to 9 months of age
- Irritable, crying, fever, diarrhea, poor appetite

Lyme disease
- Borrelia burgdorferi spirochete (carried by ticks)
- Bite and skin rash - circular red rash that continues to grow – 1 day to 1 month after being infected
- Signs/symptoms: Center of the rash may clear as it grows, giving the appearance of a bulls-eye – headache, meningitis, unilateral/bilateral facial paralysis, paralysis/ataxia.

Myasthenia Gravis
- Autoimmune neuromuscular disease involving lower motor neurons
- Weakness of voluntary muscles with repetitive activity (not enough acetylcholine for effective contraction)
- Immune system attacks synaptic junction – inflammatory response destroys acetylcholine receptors

- 20 to 30 years of age or late middle age
- Abnormal fatigue of ocular, facial, respiration muscles

Guillain-Barre' Syndrome
- AKA acute idiopathic polyneuritis
- Follows a non-specific infection that occurred 10 to 14 days earlier
- Widespread demyelization/inflammation of ascending/descending nerves
- Weakness (with sensation preserved) that increases in severity over days or weeks
- Motor paralysis and respiratory failure may result
- 85% of patients have full recovery

RETT Syndrome
- Progressive encephalopathy of unknown cause that develops in girls between 6 and 18 months
- Loss of voluntary hand movement, hand wringing movements, rigidity of the legs, growth retardation, seizures, loss of facial expression
- Deceleration of head growth between 5 and 48 months of age
- Believed to account for a large number of mental retardation children

Parkinson's disease
- Degenerative neurological disorder of the brain's dopamine neuronal systems.
  - 50 yrs. of age most commonly affected
- Symptoms (unilateral initially) – tremors at rest and with fatigue; disappearing with intended movement and sleep.
- Motor impairment causes masked facial expressions and poor blink reflex
- Behavioral change and dementia in 10 to 15% of patients

NPH (Normal Pressure Hydrocephalus)
- Noncommunicating hydrocephalus – dilated ventricles, but normal ICP
- Gait disorder, psychomotor slowing, and incontinence
- Progressive dementia with memory loss
- Some patients have a prior history of SAH, TBI, or meningitis

- Condition is correctable with VP shunt and medications

Pediatric Emergencies

Pediatric Considerations

Growth and Development

Anatomical Differences

Functional Differences

Airway Management
- Supplemental O2 to prevent intubation
- RA & "Binkies"
- Blow-by/ Nasal Cannula/ FM/ NRB

Anatomical differences in the airway
- Straight/Curved Blade
- ETT Size (16+Age/4)
- Cuffed vs. Uncuffed
- Stylet?
- 1x, 2x, 3x the tube size

Airway Management

Confirming the tube
1. Chest Rise / TV
2. Direct Visualization/ Physical Exam
3. Skin Color, SaO2
4. ETCO2 (numeric and colormetric)
5. Auscultation
6. Neg Belly Sounds
7. Port CXR   (Location on A&P Film?)
8. ABG's
9. Mist, etc, etc

Securing the tube
- Commercial securing devices
- "Pore" Tapes – Transpore / Durapore
- "Good" Tapes – Cloth / Elastoplast

Advanced airway procedures
Surgical vs. Needle Crichothyroidotomy

Reactive Airway Disease
- Differential Diagnosis
- RSV Bronchiolitis, pneumonia, croup, TB
- CF, CHF, GE reflux, tumors
- Anaphylaxis

- PE, COPD
- Foreign body aspiration

What to Know & What to Look for

Signs of Respiratory Failure
- Signs of fatigue
- Accessory muscle use
- Diaphoresis
- Inability to speak in complete sentences
- Upright/ Tripod
- O2 sat < 90% / Central Cyanosis
- Silent Chest

History / Questions to ask
- Any treatments prior to arrival?
- Ever been intubated?

Treatments
- Beta 2 Adrenergic (Albuterol, Proventil, Xopenex)
- Steroids – IV, PO or inhaled – 4-6 hours onset with PO or IV, inhaled-not for acute phase. Pedia-Pred
- Anticholinergic Bronchodilators (Atrovent)
- Judicious IV fluids
- IV Aminophylline – prevents reoccurrence, not for acute phase
- Magnesium sulfate-relieves bronchospasm
- Heliox – Gets through the mucous easier- Helium & Oxygen. Oxygen is attached to Nitrogen, too heavy to get through the mucous. (70/30 or 60/40 mix)
- IV Ketamine – sedative & Bronchodilators – drug of choice in asthma.
- Brethine / Terbutaline
- Intubation and sedation – CO2 rises in patients who are worn out.

Epiglottitis

- Inflammation and swelling of the epiglottis and surrounding tissues
- H. Influenza with bacterial component
- Hib vaccines have made this a rare occurrence.

- Can cause complete airway obstruction. Precipitated by gag reflex stimulation.
- Avoid examining the upper airway
- Suctioning

Clinical Presentation:
- Symptoms occur rapidly, causing parents to seek medical attention within 24 hrs
- Muffled voice
- Fever
- Stridor
- Labor breathing (supraglotic edema)
- Drooling
- Usually anxious
- Tripod position

Rapid recognition and treatment of airway obstruction

Lateral neck X-ray (can r/o stridor and resp distress, + thumbprint sign) **you WILL see this again!**

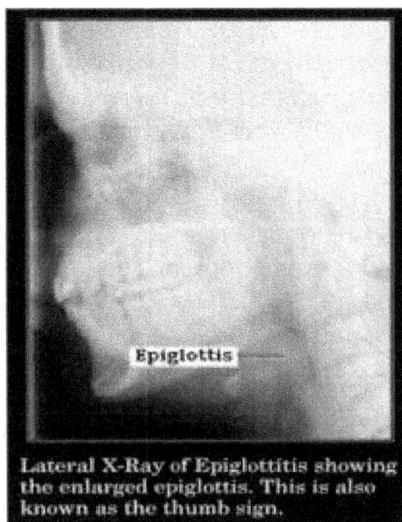

Lateral X-Ray of Epiglottitis showing the enlarged epiglottis. This is also known as the thumb sign.

If intubation is needed, Anesthesia or ENT staff if possible OR w/ inhalation induction and access to tracheotomy equipment if available

Antibiotics included, cefuroxime, cefotaxime, & ceftriaxone

Require an intensive admit!

Croup
- Viral infection that affects the larynx but can extend into trachea and bronchus
- History of fever and coryza

- As it progresses, Stridor and "barking cough"
- Rhonchi and wheezing if extends into bronchi
- Lateral neck X-Ray to r/o Epiglottitis
- Steeple Sign on PCXR. *You WILL see this again!*

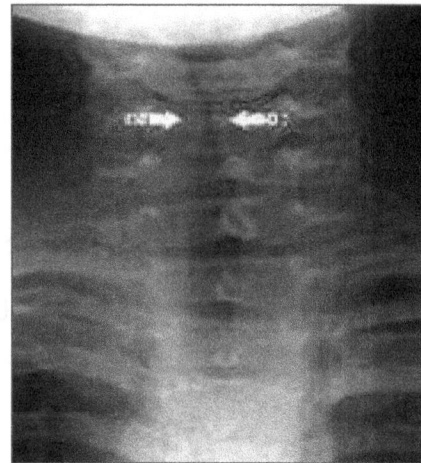

Treatment:
- Primarily supportive
- Treatment centers on dehydration and treatment of resp distress
- In rare occurrences, upper airway edema or obstruction may require intubation
- Racemic Epi, Aerosols, Dexamethasone, Prednisone

Management
- If you make them cry, they will die!!
- Avoid stimulation of the airway
- Provide humidified oxygen
- NO IV or Blood draws
- Utilize parents for comfort, if appropriate
- Surgical intervention immediately
- Endotracheal tube one size smaller than normal
- Needle cricothyrotomy
- Blood cultures - antibiotics

Head Injuries
- Big Head, Little Body Syndrome
- Open fontanel's & a larger head surface
- 80% of peds with severe CHI will have Increased ICP and cerebral edema
- 400x greater chance of having a brain injury if skull is fractured than adults

Assessment

- Level of Consciousness
- AVPU/ Pediatric GCS
- Intubation / Swelling / Paralysis / Sedatives
- Fontanel's
- Pupils
- Extremities
- Vital Signs
- HR / RR / BP / Temp

Management
- C- ABC's
- Hyperventilate?
- GCS < 8 , GCS drops by 3 or more, blown pupil, loss of extremity function
- Head Position
- Suctioning
- ? Pretreat with Lidocaine
- IVF
- NS / LR / Hypertonic NS / NO D5W

Management
- Blood Pressure Maintenance
- CPP = MAP – ICP
- MAP of 60 is needed to perfuse the kidneys *(Yep...you'll see this one too)*
- Low BP is better than High BP
- Ideal is maintaining around 90 systolic
- Pressors
- Dopamine / Neosynephrine
- If ICP is higher than BP, brain death quickly ensues.

Seizures
- Flicker effect
- Valium
- Diuretics
- Bolus vs. Continuous Infusion
- Steroids
- Not utilized in the acute phase

Spinal Cord Injuries
- Pediatric Variants
- Fulcrum of motion is at C2-C3 versus C5-C6 in adults
- Facet joints of C1-C3 are horizontal
- MOI
- 35-50% - MVC
- 30-35% - Diving/Falls
- 20-25% - Sports Injuries

Spinal Cord Injuries

- < 8 yrs old have higher cord injuries (C1-C4)
- 8 – 12 years – transitional stage
- >12 years – injury patterns like adults

SCIWORA
- Accounts for 40% of all pediatric spinal cord injuries
- Hypermobility & lax ligaments allow bones to pop out & back in.
- Disrupts microvascular blood supply
- MRI
- Crashing & Coding Kids
- Tube size calculations
- Rapid Assessment

Neuro Assessment
1. LOC/ GCS
2. Muscle tone/ Activity
3. Fontanel's
4. Respiratory
5. Breath Sounds
6. RR
7. Retractions/ Accessory muscle usage
8. Grunting / Speech
9. Cyanosis
10. Crashing – con't

Cardiovascular
- LOC
- Peripheral Pulses/ Cap refill
- Heart Rate

Rhythm (New AHA Guidelines)
- Too fast
- Too slow
- Not at all

GI/GU
- Abdominal distention/ Liver size/ Urine Output
- Crashing – con't

Skin
- Color/ Bruising
- Temperature/ Moisture
- Vital Signs
- "Normal" values
- Broselow Tapes – updates in November
- Coding Kids
- CPR Changes-Basic
- 30:2 –everyone

- 120/ minute – between the nipples
- Thumb encircling technique
- No mouth to mouth
- No one was doing it anyway
- YUCK!
- Only about 16% O2 – worsens the acidosis

Coding Kids
- Electrical Therapies
- "Sometimes all you need is a good paddling"
- 85% of arrests are V. Fib
- Defibrillation / Unsynchronized
- 2J/kg –first time x 1 shock (New guidelines vs. old)
- 2 years or older – adult paddles
- Synchronized Cardioversion
- .5-1J/kg first time
- 4J/kg second & subsequent times

Drug Therapies

Two kinds of patients
- Sick
- Really Sick

Three kinds of rhythms
- Too Fast
- Too Slow
- Not There

ECG 101
- QRS complexes are the "big blips" on the screen.
- Skinny blips originate in the atria
- Fat blips originate in the ventricle
- Lots of blips – tachycardia
- Not a lot of blips – bradycardia
- No blips- you're dead
- Blips but no pulse – you're still dead

Bradycardias
- Not enough blips
- Oxygen
- Atropine?
- Pacing
- Epinephrine
- Isuprel

Tachycardia's
- Fast & Skinny Blips
- Sinus Tachycardia

- < 200 bpm
- Oxygen, fluids, pain management

Supraventricular Tachycardia
- 200 bpm
- Sick?
- Really sick?
- Oxygen and IV fluids
- Tachycardia's
- Vagal maneuvers for kids?
- Carotid massage – not anymore
- Facial ice water immersion, occluded straw
- "Bear down"

Medications
- Atropine
- Amiodarone
- Lanoxin
- Beta Blockers

More Tachycardia
- Fast & Fat Blips
- SVT vs. VT

Ventricular Tachycardia
- Oxygen and Fluids
- Sick?
- Really sick?
- Drug Therapies
- Amiodarone
- Lidocaine
- Procainamide
- Bretylium – gone from the protocol

"Torsades de Pointes"
- Sick?
- Really sick?
- Magnesium Sulfate, Mag levels on Chemistry
- Avoid the "ide" meds

V. Fib
- Now only one shock, 2 j/kg, then the others are 4j/kg
- AHA 2005 Guidelines state only one shock, then 2 min CPR
- Advanced Airway, IV, Meds, then shock again
- Meds prepare the heart to be shocked
- Epinephrine (No more high dose)
- No more drugs down the tube
- Amiodarone

- Lidocaine

Pulseless Electrical Activity
- Find and treat cause
- Treatable vs. Non-treatable
- Epinephrine
- Atropine

Rule outs for Asystole and PEA
also treat! (MATCHED)

Myocardial Infarction

Acidoses (Respiratory and Metabolic)

Tension Pneumothorax

Cardiac Tamponade

H Hypoxemia
Hypovolemia
Hypothermia
Hypoglycemia
Hypo / Hyperkalemia

Embolus (Pulmonary)

Drug Overdose

Asystole
- Rule out fine V. Fib
- Transcutaneous Pacing (if witnessed)
- Epinephrine
- Atropine

Fluid Resuscitation
- Peripheral IV access – 17% success rate
- Intra-osseous –80%-100% success rate
- 20cc/kg fluid boluses – NS/ LR
- 10cc/kg blood products
- NS preferred / LR Ok
- No D5 anything
- Any Pressors needed?

Near -Drowning

Physiology of Drowning
- Struggling to get to the surface
- Breath holding, air panic & hunger

- Laryngospasm and glottic closure = NO air entering
- Progressive hypoxia, swallowing of water and vomiting
- Further hypoxia, muscle weakness, larynx relaxes, water into lungs
- Pronounced hypoxia- cardiac arrest

Why Doesn't Everybody Drown?
- Divers or Mammalian Reflex Theory
- Body's response to sudden asphyxia
- Profound peripheral vasoconstriction & slowing of cardiac and respiratory systems.
- Only 15% of humans have this reflex.
- Why Doesn't Everybody Drown?

Hypothermia theories
- Brain vs. Body hypothermia
- External hypothermia
- Cerebral temperature must drop by 7C within the first 10 minutes of submersion for cerebral protective effect
- Cerebral Temp only falls by a maximum of 2.5C from external hypothermia
- Internal hypothermia
- Pulmonary aspiration
- GI tract aspiration – 20% of body mass must be ingested

Management
- History of event
- What made them get in and stay in the water?
- Who was watching the child? What were they doing?
- Anatomical differences
- Any concurrent injuries present?
- Seizures / Alcohol/ Cardiac Hx
- Submersion time
- *Clean vs. Dirty water / Warm vs. Cold
- Interventions prior to EMS arrival
- Initial findings upon arrival
- C ABC's
- Oxygen
- Nebulizers for Bronchospasm
- Diuretics for pulmonary edema management
- Seizure medications
- Treat hypothermia and hypoglycemia

Outcomes
- Extremely poor prognosis for:

- 25 minute submission
- 25 minutes without heartbeat
- No pulse on arrival to ER
- GCS of 3 in ER
- Arrest
- Pupils fixed and dilated

Bites & Stings
- Incredibly exciting review of anaphylaxis physiology
- Antibody – antigen reaction
- Histamine & other substances released
- Increased capillary permeability
- Shock due to relative hypovolemia
- Intractable Bronchospasm

Causes:
- Drugs
- Diagnostic agents
- Biologic agents
- Foods
- Bites & Stings

Signs & Symptoms
- Anxiety, HA, sense of impending doom
- Hypotension, Tachycardia, Shock, Capillary leaks, Arrhythmias
- Dyspnea, Bronchospasm, Wheezing, Stridor, Obstruction
- Hematemesis, Hematochezia, Abd. Pain and cramping
- Rash, Swelling, Itching

Management

1. Early appropriate airway management
2. IVF boluses – 20cc/kg
3. Dopamine
4. Epi SQ – Not sick
5. Epi IV bolus or infusion – Sick
6. H1 Blockers – Benadryl PO or IV
7. H2 blockers – Zantac, Tagamet
8. Steroids – PO or IV
9. Bronchodilators
10. Glucagon IV

Neonatal Emergencies

Neonates
- Newborn babies, whether small or large, premature or mature
- Babies less than 28 days of age (product of term delivery)
- Neonate-31 weeks
- Neonate-35 weeks
- Neonate-Full-term

Considerations
- Circulation
- Respiration
- Thermoregulation
- Fluids and electrolytes

Circulation
- Low-resistance parallel circulation of the fetus to higher systemic resistance series circulation of the infant
- With respiration, the removal of the placenta, the fall in PVR and addition of the higher systemic resistance > permanent closure of the ductus venosus, ductus arteriosus, and foramen ovale.

Neonate-IUGR-41 weeks

Respiration
- Lungs transfer from fluid-filled state to gas exchange organ
- Initiated by cold, tactile stimulation and chemoreceptor's
- Surfactant Administration
- Thermoregulation
- Increased body surface in relation to body mass
- Less adipose tissue (brown fat)
- Heat loss through: radiation, evaporation, conduction and convection

Neonate-Acrocyanosis
- Fluids/Electrolytes
- Postnatal-Normal contraction of the extracellular fluid compartment-maintain serum electrolytes while allowing the loss of extracellular water

- After postnatal-Replace fluid and electrolyte losses and maintain fluid balance

Proper team configuration
- Mode of transport
- Documentation
- Quality improvement
- Family Support
- Isolette
- Airway Management/Oxygen
- Thermoregulation
- Fluid administration
- Wt/kg x fluid intake over 24 hrs = ml/hr
- 24

Scarf Sign

Sole Sign

Trauma
- C-spine with Airway
- Washcloth roll, v-pad, trauma dressing
- Breathing
- Blow by, cannula
- Circulation
- Scalp vein, peripheral vein
- Disability
- Pediatric Glasgow Coma Scale
- Expose
- Thermoregulation

Head Bumps

Neonate-Skin Signs

Anomalies

Neonatal Transport

Ocular, Muscolosketal, Integumentary, etc

Cranial Nerve Review

II – Optic
- Vision

III – Oculomotor
- Pupil size, most eye movements

IV – Troclear
- Eye movements

VI – Abducens
- Lateral eye movements

Eye Exam
- Follow finger across midline?
- Eyes pointing the same way? (disconjugate gaze)
- Nystagmus – eyes move in jerky movements
- Lateral – narcotics
- Ptosis – paralysis of 3rd cranial nerve

Eye Exam
- Opacity – cataracts
- Arcus senilus – light ring around cornea – normal with old age
- Ocular motility – follow finger or pen to all six fields

Eye Exam
- Visual fields
- Stand right in front of patient
- Ask pt to cover one eye
- Hold one finger in front of pt's eye
- Have pt focus on that finger
- Move finger from other hand to side
- Have pt tell you when can no longer see finger at side
- Should go as far as ear

PERRLA
- Equal?
- If awake, unequal pupils are not caused by brain injury.
- Round?
- Prior surgery? Congenital?
- Reactive to light?

- Direct – brisk, to 2 mm
- Consensual – shine light into one eye, look at pupil reaction in the other

Accommodation
- Pt looks at a distant spot, then at your finger about 5 inches from face. Pupil should constrict

Light Reflex Test

Specific Ocular injuries
- Annually, over 2.5 million Americans suffer an eye injury
- More than half a million blinding injuries occur every year
- Most commonly occurs in young male adults, followed by the elderly
- Profound social implications regarding the lost productivity
- Requirement of caring facilities and rehabilitation for the elderly

Specific Ocular injuries

- Injuries range from very mild, non-sight threatening to extremely serious with potentially blinding consequences
- Majority of injuries are minor, such as corneal abrasions or superficial corneal foreign bodies.
- Only 2-3% of all eye injuries require hospital admission
- The incidence of ocular trauma requiring hospital admission is 8.14 per 100 000 of the population annually.
- Over 10% of these people will lose useful vision in the injured eye.

Blunt injuries
- Ranges from simple "black eye" to more serious intra-ocular disruption, including rupture of the globe.
- Results in contusional and tearing damage
- Increases in orbital pressures
- Restriction in eye movements and diplopia
- Very difficult to treat effectively

Blow out fracture, with contents herniating

Le Fort Fracture

Penetrating Injuries
- Poorer prognosis than blunt injuries
- extent of damage depends on where and how far the object enters the eye
- Stabilize in place, bulky dressing
- Cover the uninjured eye as well
- NEVER remove

Burns
- Burns caused by strong acids or alkalis are among the most urgent of ophthalmic emergencies
- Severity depends on the concentration and pH of the agent and duration of exposure

Alkalis are the worst
- damage cells and penetrate the tissues rapidly

Acid injuries tend to be less severe
- they remain confined to the ocular surface

Burn management
- every second counts
- copious irrigation with water may improve the prognosis considerably
- Try and determine the affected agent
- Intra-ocular pressure raises after trauma, and causes further corneal and optic nerve damage

Le Fort I
- The result is a "floating palate" with mobility of tooth bearing segment of upper jaw
- Disturbed occlusion
- Palpable crepitation in upper buccal sulcus
- 'cracked pot' percussion note from upper teeth

Le Fort II
- produces a separation and mobility of the midface
- Gagging on posterior teeth
- Anterior open bite
- There may be diplopia and /or subconjunctival hemorrhage
- There may be Infra-orbital nerve damage

Le Fort III
- Mobile middle third of face
- Similar symptoms as Le Fort II
- There may be CSF Rhinorrhea
- 25-50% of II and III fractures
- Disturbance in smell and taste may occur
- Maintain a high index of suspicion for all facial trauma

Complications
- Extensive hemorrhage
- Airway obstruction
- Surgical airway often times needed

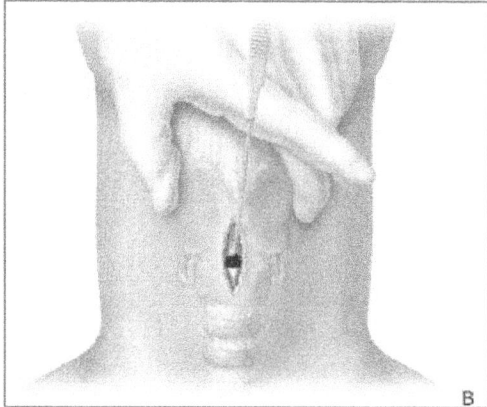

- Infection & Sinusitis
- Nerve damage (smell and taste as a clue)
- CSF leak

Ocular Problems
- 17-25% of patients will suffer some type of ocular problem
- 1-2% of patients can have blindness, diplopia, blurred vision, or future lacrimal drainage problems.

SUBJECTIVE:
- HPI/CC
- Pain: OPQRST
- Aggrav / allev
- Pre-hospital Tx
- Prior episodes

PMHx: Allergies, Meds, Immunize, Surgeries, Hospital admissions
Social: Work, tobacco, ETOH, Drugs

OBJECTIVE
- Physical Exam
- Area of concern
- Head to toe
- Diagnostics
- Nsg/Collaborative Dx

Interventions
- Medications
- Pt Education

SELECETD CONDITIONS
1. Abrasion
2. Amputation
3. Avulsion
4. Bite/Sting
5. Carpal Tunnel Syndrome
6. Compartment Syndrome
7. Contusion
8. Costochondritis
9. Foreign Bodies
10. Fracture/Dislocation
11. Bursitis/tendonitis/gout
12. Joint effusion
13. Laceration
14. Low Back Pain
15. Missile injury
16. Osteomyelitis
17. Puncture wounds
18. Strains
19. Sprains

ABRASIONS
- Injury d/t dermis/epidermis
- Falls, scrapes
- Good wound care essential to avoid "tattooing effect" from debris
- Should be cleaned in 6-8 hours
- Diabetics, Immune compromised
- Wound Cx's?
- Clean, Debride; Betadine, H2O2, NS
- Pain meds
- Tetanus
- Non stick dressing, porous
- Avoid tunnel vision
- ABCDs
- Without cooling
- Digits 12 hours
- Others 6 hours
- With cooling
- Digits poss. up 24 hours
- Proximal to wrist 12 hours

AMPUTATIONS
- Healing/reattachment points
- Children
- >50 y.o. –PVD
- Functional vs hindrance
- Can they complete rehab?
- Important: note time of injury
- Control bleeding first
- Simple dressing up to tourniquet

AMPUTATIONS
- Splint, elevate
- Copious irrigation; NS only
- Wrap stump with NS gauze
- Amputation – cover dry sterile gauze
- Place in bag
- Place that bag into a bag of ice
- Pain meds, Abx, tetanus
- Possible transfer out

AVULSIONS
- Can't approximate tissue edges
- Common in fingers
- Hemostasis sometimes difficult
- Small ones: delayed secondary healing
- Large: surgical
- Sterile NS dressing
- Vaseline/xeroform good
- Consider metal/protective splint

BITES/STINGS
- May cause
- Punctures/lacerations
- Anaphylaxis, infection, disease transmission

Sources:
- Dogs Cats
- Snakes
- Insects/Spiders
- Ticks
- Aquatics
- Human

Dogs/Cats
- Lacerations/puncture
- Possible crush injury (dog)
- 90% of ER bites are dogs
- Cat bites hi-risk infection (pasturella)
- Dogs/cats – not likely rabid
- Consider rabid: bats, foxes, raccoons
- Known vs unknown animal
- Provoked?
- Gen appearance

Dogs/Cats
- Thoroughly wash, irrigate
- Sutures +/
- Pain meds/sedation
- Abx/tetanus
- Rabies prophy

- RIG immediate protection

Snakes
- Thousands annually
- Only 10-15 mortal

Two families
- Viperidae (vipers, rattlesnakes)
- Elapidae (coral snake)
- Eastern half of the US
- Copperhead, Cottonmouth, Rattlesnake, Coral Snake

Snakes
- Hands, feet, face injuries
- 20% "dry" bites
- Croataline envenom:
- Rapid onset burning pain and edema
- Ecchymosis ensues and spreads
- Venom enzymes cause
- Local tissue damage, edema
- Hypotension, coagulopathy, shock, death

Elapidae envenom
- Little or no local reaction
- Sx onset possible delayed up to 13 hours
- Pt w/paresthesias, burning at local site at higher risk for NMB and resp compromise

Snake identification
- Pit viper: triangular head, pit between eyes, elliptical pupils
- Coral snakes: red, yellow, black bands; black head; "red on yellow kill a fellow"

Bite location
- Time
- S/Sx: weakness, paresthesias, diplopia, muscle pain, N/V/D, CP/SOB
- Pt's weight
- Pt Hx
- Previous Antivenom?
- Pit viper: fang marks
- Coral snake: tiny punctures, "chew"
- Both: edema, ecchymosis, vesicles
- Coags, FDP, FSP, U/A-myoglobin

Snakes
- Elevate slightly, immobilize
- Cardiac, Sp02

nostril

heat pit

ellipitical pupil

Timber Rattle Snake
amelanistic
(venomous)

SNAKESandFROGS.com

- Routine wound care
- Controversy over cooling
- Antivenom: give within 4 hours of bite
- Dose depends on degree of envenom
- CroFab requires no skin testing

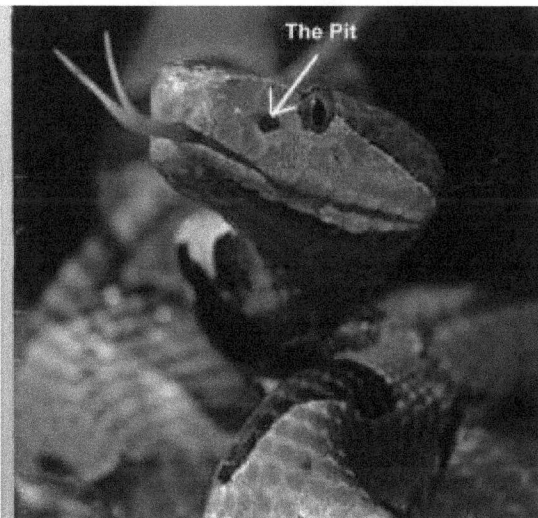

The Pit

Insects
- Bees, wasps, ants
- Concern: anaphylaxis
- Oral facial edema, respiratory, urticaria
- Hx Asthma?
- Prior sting allergy/reaction?
- Stridor, cough, hoarseness?
- Stinger still in?

Insects
- Remove stinger w/scraping motion

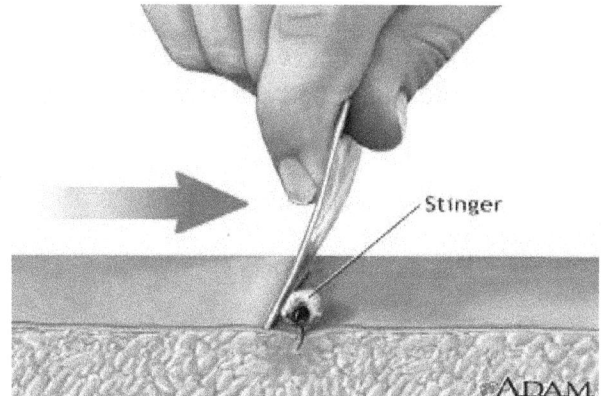

Stinger

ADAM

- Ice, baking soda, aloe vera to area
- Epi
- Benadryl
- Zantac
- Steroids
- Analgesics
- Abx?
- Tetanus

BITES & STINGS
- Pt education
- Epi-Pen
- Prevention: insect repellent

CARPAL TUNNEL SYNDROME
- Nerve entrapment
- Median nerve
- Women: men 2:1
- Pregnant women – fluid retention
- Repetitive motions
- Pain, paresthesias median nerve area

CARPAL TUNNEL SYNDROME
Tinel's Sign: percussion/tapping on median nerve; if tingling felt in hand = diagnostic

Phalen's test: forced flexion for 1 minute causes exac. of paresthesia along median nerve = diagnostic

- X-rays r/o others
- EMG Gold Standard
- Ice/Elevate/NSAIDs
- Wrist splint(s)
- Ortho, Occupational Med Referral

- Trauma, Crush injuries, High pressure injections burns
- Lower leg, forearm
- Also d/t splints, casts, PASG
- First affects low-flow system
- Later arterial flow and nerves affected

5 P's
- Pain, Pallor, Paresthesias, Paralysis, Pulse
- PMHx: Hemophilia?

COMPARTMENT SYNDROME
- U/A myoglobin
- Serum MB, CPK, CBC, Coags
- Doppler U/S?
- Compartment pressure monitoring
- 10 mmhg or less: normal/OK
- 30-40 mmhg: possible fasciotomy
- Elevate slightly, don't use Ice
- Pain meds

CONTUSIONS/BRUISES/HEMATOMA

- A closed wound with ruptured blood vessels
- Sx: swelling, discoloration, tenderness
- PMHx: check for bleeding disorders or meds that prolong bleeding
- Labs?
- Possible x-rays
- Ice/elevate/splint

COSTOCHONDRITIS
- Inflammatory chest wall pain
- Poss. r/t exertion/strain
- Pain similar to AMI or rib Fx's
- Pain: desc sharp, pleuritic
- MUST: consider CAD/AMI risk factors and r/o as indicated
- Don't forget PE, Pneumonia
- Tx: NSAIDs, Muscle relaxers, Narcs, Heat

SPRAINS/STRAINS (Know these differences)

Sprain: Ligament
Strain: Muscle/tendon

Risk factors:
- "weekend warriors"

- RA, Steroid use/injections
- Diagnostics: Ottawa ankle, knee, foot rules
- Achilles rupture: "hit in back of leg" – classic
- Swelling vs deformity

R.I.C.E.
- Immobilization
- Wt bearing limitations
- NSAIDs, Narcs

FRACTURES

DOCTOR
- D: disp vs non-disp
- O: open vs closed
- C: complete vs incomplete
- T: transverse vs linear
- O/R: Open Reduction vs closed
- Open Fx: hi-risk infection
- Crush injuries: complicated fx's

1. Impacted
2. Spiral
3. Transverse
4. Oblique
5. Complete
6. Comminuted

BEWARE: Tunnel vision!
- Splint in position found
- Ice/elevate
- Pain meds
- Possible sedation/reduction

MAST/PASG
- "considered if" intra-abdominal /pelvic bleeding with hypotension
- ACOS recommends for them for bleeding from pelvic and lower extremity Fx's

Open Fx's:
- Cover w/sterile NS dsgs
- Clean/irrigate only w/NS
- Wound mgmt
- Possible cultures
- Abx/tetanus

FRACTURES

Traction Splints
- Used to stabilize a mid-femur fraction along it's long axis using counter-traction

Without counter-traction it is possible for the spasming muscles or deformity to compromise distal NVS/CMS

Traction Splints
Indications
- For use in mid-femur fracture

Contraindications
- Concurrent knee, hip or pelvis injury
- Lower leg or ankle injury

Application
- Stabilize fracture site, manual traction
- Secure splint proximally (ischial tuberosity)
- Secure splint distally (ankle)
- Apply mechanical traction
- Re-assess distal CMS/NVS

DISLOCATIONS
- Considered an EMERGENCY d/t potential nerve damage, blood vessel damage or tissue ischemia

KNEE: (not patella) immediate reduction
ANKLE: usually needs surgery; prompt redxn

SHOULDER: needs prompt reduction
- Possible AVN
- Consider seizure cause

RADIAL HEAD -Nursemaid's elbow
- d/t sudden jerk or pulling
- Refuses to use arm
- Can flex and extend elbow, won't pronate/supinate
- Possibly no deformity and no significant pain
- May be recurrent up to 5 years old

OSTEOMYELITIS
- Infection of the bone and tissues
- D/T open wounds, puncture wounds, surgery
- Exogenous vs hematogenous sources
- Hematogenous: skin abscess, OM, UTI, Pneumonia, abscessed teeth
- Staph Aureus
- Pain, fever, malaise
- Swelling, redness, warmth
- Possible Sx of Sepsis

- PMHx: Diabetes, sickle cell, Immune compromised
- Hx of surgical procedure, fixation
- IVDU
- CBC, ESR, CRP, BMP, Blood cultures
- Bone Scan
- Parenteral IV Abx long term
- Transmission precautions
- Pain meds
- Orthopedics f/u

PUNCTURE WOUNDS
Commonly
- Stepping on nails, tacks, needles, broken glass
- Usually minimal bleeding and seal off – increasing infection risk
- Near joints esp. hi-risk
- Plantar surface through shoe inc risk of Pseudomonas d/t foreign body
- CBC, BMP, X-rays

PUNCTURE WOUNDS
- Remove FB if present
- Assist with opening, debriding, irrigating and packing if contaminated
- Abx, tetanus, pain meds
- Elevate, limit use
- Home care: warm soapy water soaks
- Should have wound check follow up made

FOREIGN BODIES
- Many things
- Wood, metal splinters, glass, clothing, GSW fragments, fishhooks etc.
- Vegetative (organic) items more likely to cause infection (thorns, wood), need to be removed

Diagnostic imaging
- X-ray vs CT
- Fluoroscopy

- Routine wound care externally
- ***Don't soak if wood embedded***
- Home care: soaks, elevate, pain meds, Abx

MISSLE INJURIES
- Penetrating: GSW, stabs and others
- D/T hi velocity, may cause bony, neurovascular and ST injuries remote from the path
- Forensic considerations; evidence

- Hi-pressure injection wounds require debridement and extensive irrigation under anesthesia
- X-rays usually
- Control local bleeding, elevation
- Routine external wound care
- Assist with debriding, exploring, irrigation

LACERATIONS

Wound mgmt goals:
- Restore function
- Repair tissue integrity
- Minimize risk of infection
- Epithelial cell growth begins as early as 6 hours post injury
- Concern about areas of flexion/extension
- Delayed closure issues

- Assess distal CMS/NVS
- Assess ROM, esp. flex/extension
- Consider x-rays
- Control bleeding
- Clean/irrigate with NS or antiseptic soap
- Debridement
- Wound closure
- Sutures, staples, Dermabond, Steri-strips
- May need splinting too

LACERATIONS

Wound care instructions
- Elevate x48 hours
- Ice
- Dressing on x48 hours

- 2-3 day wound check, first dressing change
- Sunblock for 6mo after sutures
- BEST prevention for infection is good, thorough wound cleansing, including hi-pressure NS irrigation
- Betadine?

Suture removal days
- Face 3-5
- Scalp 5-8

Upper extrem
- Non joint: 7-10
- Joint: 10-12

Lower extrem
- Thigh 7-10
- Knee 12-14
- Lower leg/foot 7-10

QUICK REVIEW

Name one Fx commonly assoc with compartment syndrome?

What are late signs of compartment syndrome?

What is initial nursing intervention for open wound near a Fx?

What is the name of inflammation of the synovial cavity surrounding a joint?

Onset of Gout – how long? rapid

Cat vs Dog bites – which likely to get infected? Cats – pasturella multicoeda

What two tests do we do to help Dx carpal tunnel? Phalen's & Tinel's

Which gives immediate protection tetanus toxoid vs tetanus immune globulin? TIG (BayTet, Hypertet)

Who needs immediate tetanus protection? Those w/no primary series

What labs might help most with Dx compartment syndrome? U/A myoglobin, CPK

What age range is a possible "red flag" for LBP? < 20 or > 50

What should you almost always r/o with costochondritis? AMI, Pneumonia, Ptx, PE, CAD

*Thank you once again; from **Critical Care Concepts**, for allowing us to assist in your training needs. We are truly honored that you allowed us into your facility! Good Luck on your exams and make sure you email me with your results when you test.* info@criticalcareconcepts.net

(Obviously not your answers or questions, as this is unethical…but what subjects you were strong in, what areas you feel we should have covered more, and any suggestions for future classes)

And lastly, your feedback is always encouraged, as this is how we grow and continue to meet your future needs as well as others.

Warmest Regards,

Richard & Christina

Review Questions

Flight Physiology

1.   Which of the following conditions is not a normal symptom of high altitude hypoxia?

      a.   Dizziness
      b.   Headache
      c.   Chest pain
      d.   Sleepiness

2.   Which of the following statements best summarizes Henry's Law?

      a.   The sum total of the partial pressures is equal to total atmospheric pressure.
      b.   The amount of gas in a solution is proportional to the partial pressure of gas above the solution.
      c.   The pressure of a gas is directly proportional to its temperature with the volume remaining constant.
      d.   At a constant temperature, a given volume of gas is inversely proportional to the pressure surrounding the gas.

3.   Decompression sickness is an example of which of the following gas laws?

      a.   Henry's Law
      b.   Boyle's Law
      c.   Dalton's Law
      d.   Graham's Law

4.   Which of the following conditions is not a result of the physiologic effect of gravitational force?

      a.   Cyanosis
      b.   Headache
      c.   Stagnant hypoxia
      d.   Blood pooling in lower extremities

5.   The physiologically deficient zone, defined as an altitude of 10,000 to 50,000 feet, is considered:

      a.   an altitude that is similar to space.
      b.   an altitude that requires pressure suits.
      c.   a zone to which the human body is adapted.
      d.   a zone in which human survival depends upon pressurization or supplemental oxygen.

6.   Which of the following statements best summarizes Dalton's Law?

      a.   The sum total of the partial pressures is equal to total atmospheric pressure.
      b.   The amount of gas in a solution is proportional to the partial pressure of gas above the solution.
      c.   The pressure of a gas is directly proportional to its temperature with the volume remaining constant.
      d.   At a constant temperature, a given volume of gas is inversely proportional to the pressure surrounding the gas.

7.   Which of the following conditions is an unavoidable stress inherent in the aviation environment?

      a.   Fatigue
      b.   Vibration
      c.   Dehydration
      d.   Hypoglycemia

8.   Which of the following statements about spatial disorientation is true?

      a.   It is caused by sound transmitted through air.
      b.   Its symptoms can mimic alcohol intoxication or extreme fatigue.
      c.   There is an inability to correctly orient oneself with respect to the horizon.
      d.   It occurs in situations where air medical team members breathe 100% oxygen at altitude.

9.   What equipment is most effective in minimizing the risk of injury from low-frequency noise?

      a.   Helmets
      b.   Earplugs
      c.   Earmuffs
      d.   Headsets

10. To increase personal resistance to fatigue, members of the air medical team can:

      a.   maintain good physical condition.
      b.   use hearing protection.
      c.   avoid dehydration.
      d.   all of the above.

11. Which of the following is not one of the four basic variables that affect gas volumetric relationships?

      a.   Temperature
      b.   Volume
      c.   The number of molecules
      d.   Altitude

12. The transport team knows that 1 liter of gas in the GI tract will increase by how much at 25,000 feet?

      a.   No increase
      b.   Two times
      c.   Three times
      d.   Five times

13. Gas exchange at a cellular level in which a gas from a higher concentration moves to a lower is an example of which gas law?

        a.   Henry's Law
        b.   Graham's Law
        c.   Boyle's Law
        d.   Dalton's Law

14. While preparing for a transport at midnight, the transport crew understands that night vision loss occurs at:

        a.   3000 feet.
        b.   5000 feet.
        c.   7000 feet.
        d.   10000 feet.

15. Which stage of hypoxia is characterized by an increase in blood pressure, heart rate, and the depth and rate of respirations?

        a.   Indifferent stage
        b.   Compensatory stage
        c.   Disturbance stage
        d.   Critical stage

16. A patient with a pneumothorax is at risk for which type of hypoxia when transported by helicopter?

        a.   Hypoxic hypoxia
        b.   Stagnant hypoxia
        c.   Hypemic hypoxia
        d.   Histotoxic hypoxia

17. Which stage of hypoxia can be caused by G forces during fixed-wing transport?

        a.   Hypoxic hypoxia
        b.   Stagnant hypoxia
        c.   Hypemic hypoxia
        d.   Histotoxic hypoxia

18. While transporting a trauma patient by fixed wing, a sudden decompression occurs at 30,000 feet. How much time before the crew loses "useful" consciousness?

        a.   15-30 seconds
        b.   30-60 seconds
        c.   90 seconds
        d.   3-5 minutes

19. After a hypoxic episode, the transport team will maintain a patients breathing at a rate of:

        a.   12-16 bpm.
        b.   12-20 bpm.
        c.   18-22 bpm.
        d.   None of the above.

20.   While completing a rotor wing transport one crewmember begins to experience pain in   their upper posterior teeth during descent. What could be the cause?

       a.   Barotitis media
       b.   Delayed ear block
       c.   Barodontalgia
       d.   Barosinusitis

21.   The transport team knows that temperature is inversely proportional to:

       a.   humidity.
       b.   oxygen.
       c.   pressure.
       d.   altitude.

22.   The human body's sensitivity to external vibration is highest between:

       a.   0.5 to 10 Hz.
       b.   0.5 to 20 Hz.
       c.   10 to 20 Hz.
       d.   5 to 11 Hz.

23.   The transport crew celebrates a birthday the night before their shift. If the crew has a hangover, what type of hypoxia are they at risk for?

       a.   Hypoxic hypoxia
       b.   Hypemic hypoxia
       c.   Stagnant hypoxia
       d.   Histotoxic hypoxia

24.   Exposure to positive acceleration can cause blackouts while unaffecting mental activity at:

       a.   2.5 Gz.
       b.   3.5 Gz.
       c.   4.5 Gz.
       d.   7.5 Gz.

25.   Manifestation of limb pain, respiratory disturbances, syncope, and skin irritation are signs of:

       a.   hypoxia.
       b.   fatigue.
       c.   decompression sickness.
       d.   dehydration.

26.   The law that explains the rise and fall of temperature in relations to gas expansion is:

       a.   Dalton's Law.
       b.   Boyle's Law.
       c.   Charles Law.
       d.   Henry's Law.

27. Which gas law, combines the properties of both Boyle's and Charles' Law?

    a.   Boyle's Law
    b.   Henry's Law
    c.   Dalton's Law
    d.   Universal Law

28. As the pressure of a gas is decreased above a liquid, the amount of gas dissolved in the liquid decreases, leading to bubble formation within the liquid. This represents which gas law?

    a.   Henry's Law
    b.   Boyle's Law
    c.   Charles Law
    d.   Dalton's Law

29. Which of the following is not a physiological effect of vibration?

    a.   Decreased ability to concentrate
    b.   Decreased motor function
    c.   Increased metabolic rate
    d.   Increased sweating

30. In a rapid decompression from 10,000 feet to 20,000 feet in a fixed-wing aircraft, this gas law explains the hypoxia experienced with increasing altitude:

    a.   Dalton's Law
    b.   Boyle's Law
    c.   Henry's Law
    d.   Graham's Law

31. The flight crew is exposed to 92 decibels (db) without protection. What is the permissible unprotected exposure time limit?

    a.   30 minutes
    b.   1 hour
    c.   2 hours
    d.   3 hours

32. During fixed-wing transport the medical crew notices that the windows are fogging and it feels cooler in the cabin. What should they consider as a problem?

    a.   Decompression
    b.   Heater not working
    c.   Normal ascent
    d.   Gravitational forces

33. During a long fixed-wing transport, the crew knows all of the following can aggravate third spacing except:

        a.   temperature changes.
        b.   hypoxia.
        c.   vibration.
        d.   G-forces.

34. During air transport by helicopter one of the crewmembers begins to get a false sense of climbing. You recognize this as which stress of flight?

        a.   Fatigue
        b.   Spatial disorientation
        c.   Hypoxia
        d.   G- forces

35. Treatment for the above includes all except:

        a.   relaxing.
        b.   applying oxygen.
        c.   allowing sensation to pass.
        d.   do not make rapid or sudden movements.

Flight Physiology Answers

1.   **c**
As altitude and hypoxia increase, a person can expect to have symptoms of apprehension, blurred or double vision, night vision decrements, dizziness, headache, sleepiness, nausea, tingling, numbness, hot and/or cold flashes, euphoria, and belligerence. Chest pain is generally not a symptom. [Krupa, D. (Ed.). (1997). Flight nursing core curriculum. Park Ridge, IL: National Flight Nurses Association. (pp. 8).]

2.   **b**
Henry's Law, also known as the Law of Gases in Solution, states that the amount of gas in solution is proportional to the partial pressure of that gas over the solution. This law can be seen in the transfer of gas between the alveoli and the blood. [Krupa, D. (Ed.). (1997). Flight nursing core curriculum. Park Ridge, IL: National Flight Nurses Association. (pp. 4).]

3.   **a**
Henry's Law is significant physiologically for decompression sickness. As a scuba diver ascends too rapidly from a deep dive, nitrogen bubbles form in the blood. [Holleran, R. (Ed.). (2003). Air and surface patient transport: Principles and practice. (3rd ed.) St. Louis: Mosby. (pp. 43).]

4.   **b**
Cyanosis is not an example of the physiologic effect of gravitational forces. Stagnant hypoxia is related to positive and negative Gz; blood pooling in the lower extremities is related to positive Gz, and headache is an effect of negative Gz. [Krupa, D. (Ed.). (1997). Flight nursing core curriculum. Park Ridge, IL: National Flight Nurses Association. (pp. 18).]

5.   **d**
The physiologically deficient zone (10,000 to 50,000 feet) is where the majority of commercial aviation occurs. A human's survival depends upon pressurization and supplemental oxygen. [Krupa, D. (Ed.). (1997). Flight nursing core curriculum. Park Ridge, IL: National Flight Nurses Association. (pp. 2-3).]

6.   **a**
Dalton's Law, also referred to as the law of partial pressure, states that the sum of the partial pressures is equal to the total atmospheric pressure. This explains hypoxia at higher altitudes. [Krupa, D. (Ed.). (1997). Flight nursing core curriculum. Park Ridge, IL: National Flight Nurses Association. (pp. 5).]

7.   **b**
Stresses inherent in the aviation environment include vibration, gravitational forces, and barometric pressure changes. Fatigue, dehydration, and hypoglycemia can be the result of the stresses of flight, but are for the most part avoidable. [Krupa, D. (Ed.). (1997). Flight nursing core curriculum. Park Ridge, IL: National Flight Nurses Association. (pp. 6-23).]

8.   **c**
Spatial disorientation is the inability to correctly orient oneself with respect to the horizon. Delayed ear block is barotrauma that occurs in situations where crew members breathe 100% oxygen at altitude and especially if continued during descent to ground level. Early symptoms of hypoxia may mimic alcohol intoxication or extreme fatigue. Sound transmitted through air causes noise stress. [Krupa, D. (Ed.). (1997). Flight nursing core curriculum. Park Ridge, IL: National Flight Nurses Association. (pp. 8, 15, 19).]

9.   **b**

Earplugs, well fitted, are the best for reduction of low frequency noise. Earmuffs, headsets and helmets are more effective for higher frequency noise, and not very effective with low frequency noise. The best protection is to use a combination when exposed to the combination of high and low frequency with high intensity noise. [Krupa, D. (Ed.). (1997). Flight nursing core curriculum. Park Ridge, IL: National Flight Nurses Association. (pp. 2).]

10.   **d**

To increase personal resistance to fatigue, the air medical team members must know personal sleep requirements and maintain them. Keep in good condition with proper diet, exercise, recreation, and moderation of smoking and alcohol. Wear personal protective gear, like hearing protection. Avoid dehydration by maintaining adequate liquid and snack intake. Limiting personal concerns prior to coming into work is also helpful to increasing personal resistance to fatigue. [Krupa, D. (Ed.). (1997). Flight nursing core curriculum. Park Ridge, IL: National Flight Nurses Association. (pp. 27).]

11.   **d**

The four basic variables that affect gas volumetric relationships are temperature, pressure, volume, and the relative mass of a gas or the number of molecules. Gas laws govern the body's physiological response to barometric pressure changes by these four variables. [Holleran, R. (Ed.). (2003). Air and surface patient transport: Principles and practice. (3rd ed.). St. Louis: Mosby. (pp. 41-43).]

12.   **c**

Gas expansion of 1 liter of volume in the GI tract will increase by 3 times at 25,000 feet. [Holleran, R. (Ed.). (2003). Air and surface patient transport: Principles and practice. (3rd ed.). St. Louis: Mosby. (pp. 51).]

13.   **b**

Graham's Law states that the rate of diffusion of a gas through a liquid medium is directly related to the solubility of the gas and inversely proportional to the square root of its density pr gram molecular weight. This means gas goes from a higher to lower concentration. [Holleran, R. (Ed.). (2003). Air and surface patient transport: Principles and practice. (3rd ed.). St. Louis: Mosby. (pp. 44).]

14.   **b**

Night vision loss occurs in the first stage of hypoxia at 5000 feet. [Holleran, R. (Ed.). (2003). Air and surface patient transport: Principles and practice. (3rd ed.). St. Louis: Mosby. (pp. 44).

15.   **b**

The second stage is the compensatory stage, which extends from 10,000 to 15,000 feet. This is the stage in which the body attempts to protect itself against hypoxia. An increase in BP, HR, RR, and respiration depth occurs. [Holleran, R. (Ed.). (2003). Air and surface patient transport: Principles and practice. (3rd ed.). St. Louis: Mosby. (pp. 44).]

16.   **a**

Hypoxic hypoxia is a deficiency in alveolar oxygen exchange. Pneumothorax is an example of a cause of reduction ion gas exchange. [Holleran, R. (Ed.). (2003). Air and surface patient transport: Principles and practice. (3rd ed.). St. Louis: Mosby. (pp. 45).]

17.   **b**

Stagnant hypoxia occurs when conditions exist that result in reduced cardiac output, pooling of the blood within certain regions of the body, a decreased blood flow to the tissues, or restriction of blood flow. [Holleran, R. (Ed.). (2003). Air and surface patient transport: Principles and practice. (3rd ed.). St. Louis: Mosby. (pp. 45).]

18.   **c**
At 30,000 feet the crew will have 90 seconds from the point of exposure to an oxygen-deficient environment to the point where deliberate function is lost. [Holleran, R. (Ed.). (2003). Air and surface patient transport: Principles and practice. (3rd ed.). St. Louis: Mosby. (pp.46).]

19.   **a**
After a hypoxic episode, the resulting hyperventilation must be controlled to achieve complete recovery, therefore the transport team will maintain a patient's breathing at a rate of 12-16 bpm. [Holleran, R. (Ed.). (2003). Air and surface patient transport: Principles and practice. (3rd ed.). St. Louis: Mosby. (pp. 47)]

20.   **d**
Direct Barodontalgia is generally manifested by moderate to severe pain that usually develops during ascent. Indirect Barodontalgia is a dull, poorly defined pain that involves the posterior maxillary teeth and develops during descent. Tooth pain during descent that involves the upper posterior teeth may be caused by Barosinusitis. [Holleran, R. (Ed.). (2003). Air and surface patient transport: Principles and practice. (3rd ed.). St. Louis: Mosby. (pp. 51).]

21.   **d**
Temperature is inversely proportional to altitude, an increase in altitude produces a decrease in temperature and, therefore a decrease in the amount of humidity. [Holleran, R. (Ed.). (2003). Air and surface patient transport: Principles and practice. (3rd ed.). St. Louis: Mosby. (pp. 43).]

22.   **b**
Research has established a human's sensitivity to external vibration is highest between 0.5 to 20 Hz because the human system absorbs most of the vibratory energy applied within this range. [Holleran, R. (Ed.). (2003). Air and surface patient transport: Principles and practice. (3rd ed.). St. Louis: Mosby. (pp. 55).]

23.   **d**
Use of alcohol can cause Histotoxic hypoxia, affect efficiency of cells to use oxygen, and interfere with metabolic activity. [Holleran, R. (Ed.). (2003). Air and surface patient transport: Principles and practice. (3rd ed.). St. Louis: Mosby. (pp. 45).]

24.   **c**
Exposure to positive acceleration usually causes deterioration of vision before causing any disturbance of consciousness. Exposure to +4.5 Gz typically produces complete loss of vision, or "blackout," while hearing and mental activity remain unaffected. [Holleran, R. (Ed.). (2003). Air and surface patient transport: Principles and practice. (3rd ed.). St. Louis: Mosby. (pp. 56).]

25.   **c**
These are classic symptoms of decompression sickness. [Holleran, R. (Ed.). (2003). Air and surface patient transport: Principles and practice. (3rd ed.). St. Louis: Mosby. (pp. 60).]

26.   **c**
Charles Law states that the volume of a gas is directly proportional to the absolute temperature of the gas, given that the mass and the pressure of the gas are constant. [Holleran, R. (Ed.). (2003). Air and surface patient transport: Principles and practice. (3rd ed.). St. Louis: Mosby (pp.43).]

27.   **d**
The Universal Gas Law is a more realistic view of gas behavior outside of controlled laboratory conditions. It combines the properties of both Boyle's and Charles Laws. [Holleran, R. (Ed.). (2003). Air and surface patient transport: Principles and practice. (3rd ed.). St. Louis: Mosby (pp. 41-44).]

28.   **a**
Henry's Law states that the quantity of gas dissolved in a liquid is directly proportional to the pressure of that gas above the liquid, provided that the gas does not react chemically within the liquid. [Holleran, R. (Ed.). (2003). Air and surface patient transport: Principles and practice. (3rd ed.). St. Louis: Mosby. (pp. 43-44).]

29.   **d**
Vibration can cause circulatory constriction and over ride the body's cooling mechanism thereby decreasing the ability to sweat. For this reason, body temperatures need to be monitored closely. [Holleran, R. (Ed.). (2003). Air and surface patient transport: Principles and practice. (3rd ed.). St. Louis: Mosby. (pp.54-55).]

30.   **c**
Henry's Law helps keep O2 dissolved in the blood. Henry's Law states that the quantity of gas dissolved in a liquid is directly proportional to the pressure of that gas above the liquid, provided that the gas does not react chemically within the liquid. [Holleran, R. (Ed.). (2003). Air and surface patient transport: Principles and practice. (3rd ed.). St. Louis: Mosby. (pp.43-44).]

31.   **c**
Two hours is the permissible unprotected exposure time limit for exposure to 92 decibels (db). [Krupa, D. (Ed.). (1997). Flight nursing core curriculum. Park Ridge, IL: National Flight Nurses Association. (pp. 42).]

32.   **a**
Physical indicators of decompression are flying debris, fogging (related to temperature drop), temperature drop, pressure decrease symptoms, and windblast. [Krupa, D. (Ed.). (1997). Flight nursing core curriculum. Park Ridge, IL: National Flight Nurses Association. (pp. 32).]

33.   b
Decreasing barometric pressure may cause leakage of intravascular space fluid into extravascular tissues. This can be aggravated by vibration, G-forces, and temperature changes. [Krupa, D. (Ed.). (1997). Flight nursing core curriculum. Park Ridge, IL: National Flight Nurses Association. (pp. 24).]

34.   b
Occulogravic illusion is a type of spatial disorientation where the individual feels a false sense of climbing. [Krupa, D. (Ed.). (1997). Flight nursing core curriculum. Park Ridge, IL: National Flight Nurses Association. (pp. 19).]

35.   b
Treatment of spatial disorientation includes having the person relax, allow sensation to subside, do not panic, do not make rapid or sudden head movements, and if it occurs to pilots they can rely on instrument. [Krupa, D. (Ed.). (1997). Flight nursing core curriculum. Park Ridge, IL: National Flight Nurses Association. (pp. 19-20).]

Primary and Secondary Assessment, Preflight Planning, and Patient Preparation

1. The first intervention to correct an inadequate airway is to:

    a. apply oxygen.
    b. insert a chest tube.
    c. reposition the head and neck.
    d. perform oral or nasal intubation.

2. Assessment of scene safety includes monitoring all of the following elements except:

    a. hazardous materials.
    b. potential for violence.
    c. mechanism of injury.
    d. landing zone security.

3. An expected outcome appropriate to a primary survey is:

    a. relief of pain.
    b. ability to sustain spontaneous respirations.
    c. relief of anxiety related to transport.
    d. appropriate fluid balance.

4. Outcomes expected with assessment and intervention in the primary and secondary assessment in an adult include all the following except:

    a. pulse rate of 60 to 100 beats per minute.
    b. capillary refill of 3 – 5 seconds.
    c. systolic BP of >90 mmHg.
    d. controlled bleeding.

5. Neurologic primary and secondary assessment can include which of the following scales:

    a. mini mental status exam.
    b. trauma score.
    c. Glasgow coma scale.
    d. none of the above.

6. What is the main purpose of the preflight walk around?

    a. To ensure that the team is familiar with the aircraft
    b. To ensure that the aircraft lights are operational
    c. To close the doors and windows of the aircraft
    d. To identify obscure hazards and ensure the security of the aircraft and equipment

7. What is the proper sequence of care of a trauma patient during a scene response?

   a. Airway management, breathing, circulation, Neurologic exam, and transport
   b. Airway management, brief Neurologic exam, patient history, safety, breathing, circulation, disability, expose, immobilization, and transport
   c. Safety while obtaining the patient history, spinal precautions with airway management, breathing, circulation, disability with brief Neurologic exam, expose, and transport
   d. Safety, patient history with spinal precautions and airway management, breathing, circulation, a brief Neurologic exam, expose, immobilize, and transport

8. Evidence preservation at a scene call includes:

   a. placing all weapons in a locked container.
   b. securing the scene by surrounding it with police tape.
   c. answering questions from the family related to the incident.
   d. cutting around bullet holes and/or knife holes in order to remove clothing.

9. Ensuring adequate oxygen levels for the patient during any flight is the responsibility of the:

   a. the pilot.
   b. the mechanic.
   c. the flight nurse on duty.
   d. the previous duty flight crew.

10. Transport of two patients concurrently requires:

   a. proper aircraft reconfiguration.
   b. adequate equipment for both patients.
   c. clear role responsibilities for the air medical personnel.
   d. all of the above.

11. Delay in leaving a facility for a patient transfer is acceptable due to all the following reasons except:

   a. bad weather occurring.
   b. airway stabilization.
   c. insertion of a primary intravenous line.
   d. complete laboratory results.

Primary/Secondary Assessment Answers
Pre-Flight Planning and Patient Preparation

1. c
A fast noninvasive technique is to reposition the head and neck. Although oxygen should be applied, it is an intervention for breathing not airway as is chest tube insertion. Intubation is an intervention to correct an inadequate airway, however noninvasive techniques that are faster should be tried first. [Krupa, D. (Ed.). (1997). Flight nursing core curriculum. Park Ridge, IL .: National Flight Nurses Association. (pp. 44-45).]

2. c
Identifying safety issues in assessing a scene includes looking for hazardous materials, landing zone security, and any potential for violence to protect the air medical personnel. The mechanism of injury is related to patient information and assessment versus the safety of the scene. [Krupa, D. (Ed.). (1997). Flight nursing core curriculum. Park Ridge, IL .: National Flight Nurses Association. (pp. 43).]

3. b
The ability to sustain spontaneous respirations is a direct outcome of care during the primary survey. The other three outcomes, although important, are not part of the primary assessment and intervention. [Krupa, D. (Ed.). (1997). Flight nursing core curriculum. Park Ridge, IL : National Flight Nurses Association. (pp. 45).]

4. b
Primary and secondary assessment and interventions have an expected outcome of a capillary refill < 2 seconds, not 3 – 5 seconds. The other three items are expected outcomes. [Krupa, D. (Ed.). (1997). Flight nursing core curriculum. Park Ridge, IL: National Flight Nurses Association. (pp. 47).]

5. c
The Glasgow Coma Scale is utilized to "score" level of consciousness and therefore is related to the primary and secondary assessment. The Mini Mental Status Exam assesses dementia and the Trauma Score is designed to look at the total body, not just Neurologic status. [Krupa, D. (Ed.). (1997). Flight nursing core curriculum. Park Ridge, IL: National Flight Nurses Association. (pp. 47).]

6. d
The purpose of the preflight walk around is to look for obscure hazards and grossly ensure the security of the aircraft and equipment. [Krupa, D. (Ed.). (1997). Flight nursing core curriculum. Park Ridge, IL: National Flight Nurses Association. (pp. 57).]

7. d
Safety must be the primary issue when arriving at a trauma scene call. Next is a history with the appropriate trauma care, including obtaining a history, applying spinal precautions and managing the airway, addressing the needs of breathing and circulation would be the next steps. [Krupa, D. (Ed.). (1997). Flight nursing core curriculum. Park Ridge, IL: National Flight Nurses Association. (pp. 59-60).]

8. d
Evidence preservation includes cutting around bullet or knife holes. [Holleran, R. (1994). Pre-hospital nursing: A collaborative approach. St. Louis: Mosby. (pp. 118).]

9. c
The flight nurse on duty is responsible for the adequacy of oxygen for the patient during any flight. [Krupa, D. (Ed.). (1997). Flight nursing core curriculum. Park Ridge, IL: National Flight Nurses Association. (pp. 58).]

10. **d**
Proper aircraft reconfiguration and clear role responsibilities for each crewmember is essential during a two patient flight. Ideally two aircraft would be requested, but this is not always available. There is no standard to require two ventilators or require an aircraft to carry blood products. [Krupa, D. (Ed.). (1997). Flight nursing core curriculum. Park Ridge, IL: National Flight Nurses Association. (pp. 61).]

11. **d**
Complete lab results can be faxed to the receiving facility and are rarely a reason for a delay in transport. Obtaining airway control and IV access can cause a brief delay. The pilot may have weather conditions that preclude flying. [Holleran, R. (Ed.). (2003). Air and surface patient transport: Principles and practice. (3rd ed.). St. Louis: Mosby. (pp.99).]

Airway Management, Oxygen Therapy, and Respiratory Transport

1. To avoid ischemic damage to the underlying tracheal mucosa, endotracheal tube cuff pressure should be no greater than:

    a. 75 mm Hg.
    b. 50 mm Hg.
    c. 25 mm Hg.
    d. 15 mmHg.

2. The flight crew is preparing to intubate a patient who has sustained burns over approximately 50% of his total body surface area, including possible airway burns. Prior to administering Succinylcholine, it is important for the flight nurse to establish the time of injury, because use of this agent in patients with burns more than 12 hours old can cause serious:

    a. hyperkalemia.
    b. hypernatremia.
    c. hyponatremia.
    d. hypercalcemia.

3. All of the following are contraindications to digital intubation except:

    a. an obese or short-necked patient.
    b. an alert or semiconscious patient.
    c. the inability to open the patient's mouth.
    d. insufficient resources to adequately suction the pharynx.

4. To avoid trauma to the soft palate in children, insert an oropharyngeal airway by positioning it:

    a. in an upright position until it is in place.
    b. in an upright position while depressing the tongue with a tongue blade.
    c. sideways until it is in place and then rotate it 90 degrees to lie on the tongue.
    d. upside down until it is in place and then rotate it 180 degrees to lie on the tongue.

5. Cricoid pressure (Selleck Maneuver) is a useful adjunct to endotracheal intubation for all of the following reasons except it:

    a. prevents aspiration in patients who are actively vomiting.
    b. allows downward pressure to bring the tracheal opening into view.
    c. prevents over inflation of the stomach during positive pressure ventilation with a mask.
    d. prevents aspiration, if the patient has fasciculation's after administration of Succinylcholine.

6. The airway of a child younger than 8 years old differs from that of an adult in which of the following ways?

    a. The narrowest portion of the child's airway is the thyroid cartilage
    b. The larynx is more anterior in the child
    c. The child's tongue is smaller
    d. None of the above

7. Pulse oximetry is a valuable tool for all of the following clinical applications except:

    a. detecting hypoxemia during airway maneuvers.
    b. being used as an adjunct to determine blood pressure.
    c. detecting hypoxemia when no other symptoms are present.
    d. detecting hypoxemia in patients with acute carbon monoxide poisoning.

8. An oxygen saturation level of 92% in an alert, asymptomatic patient would be considered:

    a. within normal limits.
    b. high if the patient is a smoker.
    c. an indication of possible hypoxemia.
    d. irrelevant because ambient light may cause inaccuracies in the sensor.

9. A 63-year-old man who has had a myocardial infarction and now possible aspiration pneumonia is being transported for definitive care. He is unresponsive and intubated. He has a blood pressure of 110/60 mm Hg, a heart rate of 110 beats/min, and respirations of 16/min. He is being mechanically ventilated and has no spontaneous respirations. Dopamine is infusing at 10mcg/kg/min.
   Current ventilator settings are as follows: tidal volume of 750 mL, FIO_2 of 60% (0.6), and an MV of 1 6/mm.

   Current arterial blood gas measurements are a pH of 7.34, a PaO_2 70 mm Hg, a $PaCO_2$ 50 mm Hg, and HCO_3 19 mEq/L. Pulse oximetry reading is 92%. The next treatment priority would be to:

    a. increase the FIO_2 and ventilator rates.
    b. wean the dopamine to 7.5 mcg/kg/min.
    c. continue transport with no additional interventions.
    d. give a fluid bolus of 250 mL of lactated Ringer's solution.

10. During transport, an intubated patient with emphysema is at risk for all of the following except:

    a. accidental extubation.
    b. loss of hypoxemic drive.
    c. hypercarbia due to equipment failure.
    d. barotrauma secondary to positive pressure ventilation.

11. The term oxygen saturation refers to the percentage of oxygen that is:

    a. bound to hemoglobin.
    b. dissolved in the plasma.
    c. also known as oxygen capacity.
    d. carried by both plasma and hemoglobin.

12. A 5-year-old boy is to be transported from a community hospital with a diagnosis of mild Epiglottitis. The referring physician has elected to withhold intubation at this time. Preparation for transport would include all of the following except:

    a. rapidly obtaining IV access.
    b. allowing his mother to fly along.
    c. administering humidified oxygen therapy.
    d. allowing the child to maintain a position of comfort.

13. The term "vital capacity" refers to the:

    a. amount of air remaining in the lungs after normal exhalation.
    b. maximum amount of air that can be inhaled with the greatest inspiratory effort.
    c. amount of air that can be inhaled beginning at the normal expiratory level until the lungs are maximally expanded.
    d. maximum amount of air that can be expelled from the lungs after a maximal inhalation and a forced exhalation.

14. A patient with carbon monoxide poisoning has blood drawn to measure arterial blood gases just before departure from the referring facility. The results indicate that he has a pH of 7.18, a PaO2 of 70 mm Hg, a PaCO2 of 40 mm Hg, a HCO3 of 19 mEq/L, and an oxygen saturation level of 100%. Based on these results, the flight crew concludes that the patient is in:

    a. respiratory alkalosis.
    b. respiratory acidosis.
    c. metabolic acidosis.
    d. metabolic alkalosis.

15. Acute respiratory failure is most commonly characterized by:

    a. initial metabolic acidosis.
    b. hypercarbia without hypoxia.
    c. hypoxia without hypocarbia.
    d. severe hypoxia accompanied by hypercarbia.

16. Objective findings characteristic of the chronic obstructive pulmonary disease patient on physical exam include all of the following except:

    a. pursed lip breathing.
    b. distant breath sounds.
    c. pale, dusky, or cyanotic skin.
    d. absence of breath sounds on one side.

17. A patient experiencing an acute episode of asthma generally requires intubation if:

    a. the pH is less than 7.3.
    b. the PaCO2 is greater than 45 mm Hg.
    c. there is a decreased level of consciousness.
    d. the PaO2 is less than 75 mm Hg, despite high-flow oxygen.

18. A patient sustains a single gunshot wound to the left lower chest near the epigastrium. The presence of a pneumothorax with particulate matter coming from the chest tube may be indicative of:

    a. gastric trauma.
    b. esophageal trauma.
    c. a diaphragmatic rupture.
    d. injury to the tracheobronchial tree.

19. A properly placed nasogastric tube in the chest cavity is most typically seen in patients with:

a. pneumothorax.
b. esophageal trauma.
c. injuries to the trachea.
d. a ruptured diaphragm.

20. What is the priority nursing diagnosis for a patient in cardiopulmonary arrest?

a. Fluid volume deficit
b. Impaired gas exchange
c. Ineffective airway clearance
d. Altered cardiopulmonary tissue perfusion

21. The flight crew should note the amount of blood obtained after initial insertion of a chest tube in a patient with a hemothorax, as this amount is:

a. recorded every 10 minutes for the first hour.
b. not as significant as the hourly output thereafter.
c. to be replaced 1:1 with packed red blood cells.
d. indicative of massive hemothorax if greater than 1500 mL.

22. Which of the following statements about adult respiratory distress syndrome or acute lung injury is true?

a. Its overall mortality rate is 90%
b. It results in cardiogenic pulmonary edema
c. It is a diffuse lung injury that develops in response to a variety of insults.
d. It is characterized on chest radiographs by a 'white out' of a unilateral lung field

23. Expected outcomes or evaluation of patients transported with pulmonary embolism include:

a. resolution of hemoptysis.
b. increase in oxygen requirements.
c. increased hemoglobin and hematocrit levels.
d. maintenance of coagulation times at previous or desired level or protocols..

Airway Management, Oxygen Therapy, and Respiratory Transport Answers

1. **c**

Cuff pressures that are greater than 25 mm Hg. may begin to cause ischemic changes to the tracheal mucosa. [Holleran, R. (Ed.). (2003). Air and surface patient transport: Principles and practice. (3rd ed). St. Louis: Mosby. (pp. 176).]

2. **a**

Use of Succinylcholine can cause life-threatening hyperkalemia in patients with burn or crush injuries over 12 hours old. It may produce a rise in serum potassium of less than 0.5 mEq/L in healthy adults. [Krupa, D. (Ed.). (1997). Flight nursing core curriculum. Park Ridge, IL: National Flight Nurses Association. (pp. 83).] . [Holleran, R. (Ed.). (2003). Air and surface patient transport: Principles and practice. (3rd ed). St. Louis: Mosby. (pp. 192).],

3. **a**

Digital intubation may be useful in the obese or short-necked patient. It is contraindicated in conscious patients due to possible injury to the operator. It is also important to have adequate suction available to minimize the risk of aspiration. [Krupa, D. (Ed.). (1997). Flight nursing core curriculum. Park Ridge, IL: National Flight Nurses Association. (pp. 75).], [Holleran, R. (Ed.). (2003). Air and surface patient transport: Principles and practice. (3rd ed). St. Louis: Mosby. (pp. 180).]

4. **b**

Insertion of the oropharyngeal airway in the upright position while using a tongue blade is the safest way to avoid trauma to the soft palate of the child. It may also minimize the risk of pushing the tongue back into the posterior oropharynx and causing an airway obstruction. [Krupa, D. (Ed.). (1997). Flight nursing core curriculum. Park Ridge, IL: National Flight Nurses Association. (pp. 72).]

5. **a**

Cricoid pressure may cause esophageal rupture if the patient begins to actively vomit. In this case, cricoid pressure should be released and the airway should be cleared with suction. Cricoid pressure has been helpful in bringing the airway of a patient with an anterior tracheal opening into view. It also helps to prevent aspiration in the patient who develops fasciculation's. Cricoid pressure also helps to prevent over inflation of the stomach during ventilation with a bag-valve-mask. [Krupa, D. (Ed.). (1997). Flight nursing core curriculum. Park Ridge, IL: National Flight Nurses Association. (pp. 83).], [Holleran, R. (Ed.). (2003). Air and surface patient transport: Principles and practice. (3rd ed). St. Louis: Mosby. (pp. 177).]

6. **b**

The child's larynx is more anterior than that of the adult. The cricoid cartilage is the narrowest portion of the airway while the tongue is larger than that of the adult. [Krupa, D. (Ed.). (1997). Flight nursing core curriculum. Park Ridge, IL: National Flight Nurses Association. (pp. 88).], [Holleran, R. (Ed.). (2003). Air and surface patient transport: Principles and practice. (3rd ed). St. Louis: Mosby. (pp. 185).]

7. d
The oxygen saturation may appear normal in the patient who has carbon monoxide poisoning despite hypoxia. The pulse oximeter can detect occult hypoxemia when no other symptoms are present. It can be used to monitor for hypoxemia during airway maneuvers. It can also be used to help determine blood pressure by inflation of the BP cuff while watching for the disappearance of the pulse oximeter reading; the blood pressure is then read on the manometer. [Krupa, D. (Ed.). (1997). Flight nursing core curriculum. Park Ridge, IL: National Flight Nurses Association. (pp. 86).], [Holleran, R. (Ed.). (2003). Air and surface patient transport: Principles and practice. (3rd ed). St. Louis: Mosby. (pp. 197-199).]

8. c
Pulse oximetry is useful in detecting early hypoxemia even if no other symptoms are present. A reading of 92% correlates with a PaO2 of approximately 70 mm Hg. [Holleran, R. (Ed.). (2003). Air and surface patient transport: Principles and practice. (3rd ed). St. Louis: Mosby. (pp. 197-198).]

9. a
The patient's blood gases indicate a respiratory acidosis as well as hypoxemia. Appropriate action would be to increase respiratory rate to decrease pCO2 and increase FiO2 to increase oxygenation. Intravenous fluid boluses could increase the risk for development of adult respiratory distress syndrome (ARDS). Weaning dopamine is not indicated at this point since he is normotensive. [Krupa, D. (Ed.). (1997). Flight nursing core curriculum. Park Ridge, IL: National Flight Nurses Association. (pp. 123).]

10. b
Normally, emphysema patients rely on a hypoxemic drive to maintain their respiratory function. Since she is being manually ventilated, she is not at risk for apnea. She is at risk for accidental extubation and equipment failure. This patient is also at risk for barotrauma due to high airway pressures and possible pulmonary blebs that could rupture leading to pneumothoraces. [Krupa, D. (Ed.). (1997). Flight nursing core curriculum. Park Ridge, IL: National Flight Nurses Association. (pp. 114).]

11. a
Oxygen saturation refers to the amount of oxygen that is bound to hemoglobin. Oxygen in the plasma is measured as PaO2 in blood gases. Oxygen capacity is a portion of the formula for determining oxygen content and is based on the hemoglobin value multiplied by 1.34. [Holleran, R. (Ed.). (2003). Air and surface patient transport: Principles and practice. (3rd ed). St. Louis: Mosby. (pp. 414-415).]

12. a
Interventions such as intravenous access may agitate the child further and cause complete obstruction of the airway. Position of comfort and allowing the mother to accompany the patient may alleviate agitation. Humidified oxygen may prevent further swelling while preventing hypoxia. [Holleran, R. (Ed.). (2003). Air and surface patient transport: Principles and practice. (3rd ed). St. Louis: Mosby. (pp. 597).]

13. d
Vital capacity is the maximum amount of air that can be expelled after maximal inhalation and exhalation. Functional residual capacity is the air remaining in the lungs after normal exhalation. Total lung capacity is the maximum amount of air that can be inhaled with the greatest inspiratory effort. Inspiratory capacity is the amount of air that can be inhaled beginning at the normal expiratory level until the lungs are maximally expanded. [Holleran, R. (Ed.). (2003). Air and surface patient transport: Principles and practice. (3rd ed). St. Louis: Mosby. (pp. 413).]

14. **c**
Metabolic acidosis [due to decreased pH and bicarbonate levels]. Patients with carbon monoxide toxicity often present with metabolic acidosis from excessive lactic acid buildup. [Krupa, D. (Ed.). (1997). Flight nursing core curriculum. Park Ridge, IL: National Flight Nurses Association. (pp. 606).]

15. **d**
 Acute respiratory failure occurs when illness or impairment of the respiratory system results in inability of the lungs to maintain adequate oxygenation of the blood. It occurs with or without impairment of ventilation. This is characterized by severe hypoxia accompanied by hypercarbia. [Krupa, D. (Ed.). (1997). Flight nursing core curriculum. Park Ridge, IL: National Flight Nurses Association. (pp. 106).]

16. **d**
Breath sounds may be diminished but are not absent on one side with COPD. Signs of hypoxia, pale, dusky or cyanotic color are frequently present. Pursed lip breathing allows the patient to provide a longer expiratory time for removal of trapped air. [Krupa, D. (Ed.). (1997). Flight nursing core curriculum . Park Ridge, IL: National Flight Nurses Association. (pp. 114).]

17. **c**
 Some asthmatic patients tolerate remarkable changes in their arterial blood gases. Intubation should be considered when they are no longer compensating and develop changes in mentation and/or adequate respiratory effort. [Krupa, D. (Ed.). (1997). Flight nursing core curriculum. Park Ridge, IL: National Flight Nurses Association. (pp. 112, 139).]

18. **b**
 Esophageal injury should be considered for patients who: 1) have a left pneumothorax or hemothorax without a rib fracture, 2) have sustained a severe blow to the lower sternum or epigastrium and are in severe pain or shock out of proportion to the injury, or 3) have particulate matter in their chest tube. [Krupa, D. (Ed.). (1997). Flight nursing core curriculum. Park Ridge, IL: National Flight Nurses Association. (pp. 135).]

19. **d**
Traumatic diaphragmatic rupture is usually the result of severe blunt injury to the lower chest, upper or lower abdomen. The left hemi diaphragm is the most common site for these tears, which allows herniation of abdominal contents into the chest cavity. When a gastric tube is placed it may show up in the chest cavity on radiographic exam. [Krupa, D. (Ed.). (1997). Flight nursing core curriculum. Park Ridge, IL: National Flight Nurses Association. (pp. 134).]

20. **d**
Cardiopulmonary arrest primary nursing diagnoses is altered cardiopulmonary tissue perfusion. [Krupa, D. (Ed.). (1997). Flight nursing core curriculum. Park Ridge, IL: National Flight Nurses Association. (pp. 101).]

21. **d**
Massive hemothorax is the accumulation of more than 1500 ml of blood in the pleural cavity. [Krupa, D. (Ed.). (1997). Flight nursing core curriculum. Park Ridge, IL: National Flight Nurses Association. (pp. 128).]

22. **c**

Adult respiratory distress syndrome (ARDS) is an acute lung injury of impaired oxygenation resulting from a variety of insults. The diffuse injury results in non-cardiogenic pulmonary edema. Predisposing factors associated with the development of ARDS are numerous and include: hypovolemic shock, pulmonary contusion, gastric aspiration, bacterial sepsis, inhaled toxins, Pancreatitis, and multiple trauma. Overall mortality of ARDS is typically greater than 50%. [Krupa, D. (Ed.). (1997). Flight nursing core curriculum. Park Ridge, IL: National Flight Nurses Association. (pp. 108).]

23. **d**

Thrombolytic therapy is indicated for patients with acute massive pulmonary embolus and those with respiratory or circulatory failure. Expected outcomes for a patient with pulmonary embolus would be decreasing clinical symptoms, normal ventilation and arterial blood gases, and coagulation times at the desired level to evaluate heparin therapy. [Krupa, D. (Ed.). (1997). Flight nursing core curriculum. Park Ridge, IL: National Flight Nurses Association. (pp. 118).]

Cardiovascular Transport

1.   A 67-year-old man with coronary artery disease is being transported in preparation for heart catheterization. In the last 15 minutes, the helicopter has ascended to 7,500 feet, and the patient reports substernal chest pressure. His oxygen saturation levels have decreased, while his respiratory and heart rates have increased. The increase in altitude could be responsible for the change in his condition due to:

   a.   a decrease in coronary artery blood flow.
   b.   an increased risk of pulmonary embolism.
   c.   a decrease in the alveolar partial pressure of oxygen.
   d.   a decrease in the oxygen carrying capacity of hemoglobin.

2.   Which of the following statements best describes the chest pain associated with unstable angina or crescendo angina?

   a.   It occurs at rest or on early rising and is almost always relieved by nitroglycerin.
   b.   It is usually precipitated by physical exertion or emotional stress and relieved by rest
   c.   It increases in frequency and duration with time and becomes unresponsive to previously effective treatments, such as rest and nitroglycerin
   d.   It is caused by a fixed obstruction and concomitant spasm of the coronary arteries, which precipitates ischemia with or without increased oxygen demand

3.   A patient who has an episode of sudden, intense substernal pain   described as "ripping" and "tearing" pain now reports that the pain has moved to his back. The blood pressure in the right arm is 177/86 mm Hg, and the blood pressure in the left arm is 92/65 mm Hg.   The patient's signs and symptoms are most likely due to:

   a.   Pericarditis.
   b.   cardiogenic shock.
   c.   acute aortic dissection.
   d.   acute myocardial infarction.

4.   Which of the following hemodynamic parameters is most indicative of cardiogenic shock?

   a.   Systolic blood pressure 80 mm Hg, cardiac index of 1 .8 L/min/m2, and PAWP of 30 mm Hg
   b.   Systolic blood pressure of 90 mm Hg, cardiac index of 2.2 L/min/m2, and PAWP of 5 mm Hg
   c.   Systolic blood pressure 120 mm Hg, cardiac index of 4 L/min/m2, and PAWP of 12mm Hg
   d.   Systolic blood pressure of 140 mm Hg, cardiac index of 3 L/min/m2, and PAWP of 8 mm Hg

5. A patient with a diagnosis of myocardial infarction is to be flown by helicopter to a nearby hospital for cardiac catheterization. The patient begins to hyperventilate and sweat as he expresses his fear of heights. Which of the following nursing diagnoses best describes these new symptoms?

    a. Fluid volume excess secondary to decreased cardiac output
    b. Fear or anxiety related to diagnosis, treatment, and prognosis
    c. Decreased cardiac output secondary to decreased myocardial contractility
    d. Altered myocardial tissue perfusion secondary to imbalance between supply and demand

6. A patient with an extensive anterior wall myocardial infarction has a blood pressure of 88/60 mm Hg, pulmonary artery wedge pressure of 8 mm Hg, and cardiac index of 2.O L/min/m2. Which of the following nursing diagnoses best describe this condition?

    a. Ventilation perfusion disorder
    b. Fluid volume excess secondary to decreased cardiac output
    c. Decreased cardiac output secondary to decreased myocardial contractility
    d. Activity intolerance secondary to imbalance between supply and demand

7. Elevated ST segments in leads V1 through V4 suggest what type of myocardial ischemia?

    a. Lateral
    b. Inferior
    c. Anteroseptal
    d. Anterolateral

8. All of the following are true about the intra-aortic balloon pump except:

    a. the balloon is inflated during diastole.
    b. balloon-pump counterpulsation decreases systemic vascular resistance.
    c. balloon-pump counterpulsation can augment cardiac output by at least 30%.
    d. balloon-pump counterpulsation improves coronary artery perfusion pressures.

9. Nursing intervention for a patient with dilated Cardiomyopathy is most likely to result in all of the following outcomes except:

    a. the patient being awake and oriented.
    b. urine output of 15 ml/hour.
    c. heart rate of between 60 and 100 beats/pm.
    d. systolic blood pressure of greater than 90 mm Hg.

10. Oxygen, nitroglycerin, and morphine sulfate are given to a patient with chest pain in order to:

    a. stabilize chest pain by providing treatment.
    b. relieve chest pain by balancing oxygen supply and demand.
    c. decrease cardiac workload by maintaining the systolic blood pressure at 90 mm Hg.
    d. improve coronary artery perfusion pressure by maintaining the heart rate over 100 beats/min.

11. Potential complications during air medical transport which are associated with temporary cardiac pacing include:

    a. failure to capture.
    b. sensing problems.
    c. cardiac tamponade.
    d. all of the above.

12. All of the following are interventions that act to decrease myocardial oxygen demand except:

    a. decreasing pain.
    b. decreasing afterload.
    c. decreasing diastolic pressure to less than 50 mm Hg.
    d. decreasing catecholamine effects on heart rate, blood pressure, and contractility.

13. Which of the following interventions is not indicated to maintain or improve the oxygenation status of patients transported with pulmonary edema?

    a. Administer analgesic
    b. Elevate the legs to increase venous return
    c. Place patient in a semi-Fowler's position
    d. Adjust Fi02 as needed for changes in altitude

14. Adequate ventricular filling pressure in a patient in cardiogenic shock is indicated by a pulmonary artery wedge pressure of:

    a. less than 5 mm Hg.
    b. 5 to 12 mm Hg.
    c. 15 to 18 mm Hg.
    d. 20 to 25 mm Hg.

15. A victim of a motorcycle crash sustains an open, angulated fracture of the left ankle with suspected vascular injury. All of the following are appropriate interventions except:

    a. pain medications, as necessary.
    b. maintaining the left ankle below the level of the heart.
    c. immobilization of the extremity to prevent further injury.
    d. frequent reassessment of pulses, sensation, and movement distal to the injury.

16. Initial management of a patient with suspected cardiac tamponade should include:

    a. intubation.
    b. Pericardiocentesis.
    c. initiation of a rapid IV fluid bolus.
    d. insertion of a mediastinal chest tube.

17. A patient with acute cardiovascular disease is transported by helicopter at an altitude of 8,000 feet. This patient is most susceptible to what type of hypoxia?

    a. Hypoxic
    b. Stagnant
    c. Hypemic
    d. Histotoxic

18. A 48-year-old man with an acute myocardial infarction is being transported when he goes into ventricular fibrillation and has no pulse. The flight crew should first:

a. ask the pilot to land so the patient can be defibrillated.
b. begin cardiopulmonary resuscitation (CPR) and defibrillate the patient once the helicopter lands.
c. begin CPR, give a bolus with Lidocaine 1 mg/kg, and then initiate a Lidocaine infusion at 2 mg/mm.
d. stay clear of the patient and stretcher while defibrillating the patient with 200 joules and inform the pilot.

19. The intervention of choice for a patient with multi vessel coronary artery occlusion is:

a. thrombolytic therapy.
b. placement of an intra-aortic balloon pump.
c. emergent coronary artery bypass graft (CABG) surgery.
d. percutaneous transluminal coronary angiography (PTCA).

20. A patient with angina is given IV nitroglycerin in order to:

a. increase venous return.
b. increase oxygen consumption.
c. decrease coronary blood flow.
d. provide smooth muscle relaxation.

21. The goal of treatment for a patient with a possible diaphragmatic rupture is to:

a. decompress the stomach.
b. maintain adequate circulation.
c. maintain adequate oxygenation.
d. insert a chest tube to allow expansion of the lung.

22. The most common cause of failure to pace is:

a. over sensing.
b. wire fracture.
c. under sensing.
d. battery depletion.

23. Which of the following findings is not an indication for emergency pacemaker therapy?

a. Overdrive of tachydysrhythmias
b. Asymptomatic bifascicular or trifascicular block
c. Symptomatic bradycardia unresponsive to drug therapy
d. Pulseless activity due to electrolyte abnormalities

24. Absolute contraindications for thrombolytic therapy include all of the following except:

a. intracranial tumor.
b. uncontrolled hypertension.
c. any surgeries longer than one year ago.
d. history of intracranial bleeding or cerebrovascular accident.

25. A flight nurse preparing a patient with existing intravascular hemodynamic monitoring lines for transport need not:

    a. secure and label all lines clearly.
    b. maintain a flush system during transport.
    c. handle the patient carefully to prevent complications.
    d. remove all intravascular invasive lines for patient safety.

26. A patient who has sustained a right ventricular myocardial infarction may exhibit which of the following hemodynamic findings?

    a. Sv02 of 65%
    b. CVP 2O mm Hg
    c. Cardiac output of 4.2 L/min
    d. None of the above

Cardiovascular Transport Answers

1. c

Hypoxia poses one of the greatest threats to a patient with coronary artery disease. Physiologic changes occur as barometric pressure at altitude decreases, causing a reduction in the alveolar partial pressure of oxygen. [Holleran, R. (Ed.). (2003). Air and surface patient transport: Principles and practice. (3rd ed.). St. Louis: Mosby. (pp. 356-357).]

2. c

Unstable angina represents a change in symptoms characterized by increasing frequency of attacks, increased duration, lack of response to previously effective treatment, or angina at rest. Unstable angina is a phrase denoting stable angina that has changed in the timing, frequency, intensity, duration and quality. [Krupa, D. (Ed.). (1997). Flight Nursing Core Curriculum. Park Ridge, IL.: National Flight Nurses Association. (pp. 151).] [Holleran, R. (Ed.). (2003). Air and surface patient transport: Principles and practice. (3rd ed.). St. Louis: Mosby. (pp. 359-360).]

3. c

Aortic dissection causes an abrupt onset of severe pain often described as tearing, ripping, stabbing or knifelike. The pain may be migratory from the origin along the area of dissection. The blood pressure may be normal or elevated, often as high as 200 mm Hg systolic, with a significant difference between both arms. [Holleran, R. (Ed.). (2003). Air and surface patient transport: Principles and practice. (3rd ed.) St. Louis: Mosby. (pp. 395-396).] [Krupa, D. (Ed.). (1997). Flight Nursing Core Curriculum. Park Ridge, IL: National Flight Nurses Association. (pp. 167).]

4. a

Hemodynamically, patients in cardiogenic shock manifest marked hypotension with systolic blood pressure less than 80 mm Hg, low cardiac index less than 1.8 L/min/m2, decreased urinary output, elevated heart rates and a pulmonary artery wedge pressure greater than 18 mm Hg. [Holleran, R. (Ed.). (2003). Air and surface patient transport: Principles and practice. (3rd ed.). St. Louis: Mosby. (pp. 377).]

5. b

Fear or anxiety related to transport, diagnosis, treatment or prognosis is a potential nursing diagnosis for any cardiovascular patient that is air or ground transported. [Krupa, D. (Ed.). (1997). Flight Nursing Core Curriculum. Park Ridge, IL: National Flight Nurses Association. (pp. 150).]

6. c

An anterior wall myocardial infarction affects the contractility of the left ventricle. In this scenario, the blood pressure and cardiac index are low, however the pulmonary artery wedge pressure is normal which indicates there isn't fluid overload at this time. [Holleran, R. (Ed.). (2003). Air and surface patient transport: Principles and practice. (3rd ed.). St. Louis: Mosby. (pp. 366-369).]

7. c

ST segment elevation in leads V1 through V4 is indicative of an anteroseptal myocardial ischemia. [Holleran, R. (Ed.). (2003). Air and surface patient transport: Principles and practice. (3rd ed.). St. Louis: Mosby. (pp. 366-369).]

8. c

Balloon-pump counter pulsation can augment the cardiac output by as much as 10% to 20%. [Holleran, R. (Ed.). (2003). Air and surface patient transport: Principles and practice. (3rd ed.) St. Louis: Mosby. (pp. 380-385).]

9. **b**
Urine output > 30 ml/hour, systolic blood pressure >90 mm Hg, normal heart rate of 60 -100, and for the patient to be awake and oriented are all expected outcomes when treating patients with dilated cardiomyopathy. [Krupa, D. (Ed.). (1997). Flight nursing core curriculum. Park Ridge, IL: National Flight Nurses Association. (pp. 162).]

10. **b**
Relief of chest pain is the expected outcome. This is achieved by balancing oxygen supply with demand. [Krupa, D. (Ed.). (1997). Flight nursing core curriculum. Park Ridge, IL: National Flight Nurses Association. (pp. 152-153).]

11. **d**
Some of the complications encountered with temporary cardiac pacing that may occur during air medical transport include, sensing problems, failure to capture, myocardial penetration, and cardiac tamponade. [Holleran, R. (Ed.). (2003). Air and surface patient transport: Principles and practice. (3rd ed.). St. Louis: Mosby. (pp. 375-376).]

12. **c**
Implemented measures to decrease myocardial oxygen demand include: 1) decreasing myocardial wall tension by decreasing preload, 2) decreasing catecholamine effects of increased heart rate, increased blood pressure and increased contractility, 3) decreasing afterload, 4) decreasing central adrenergic output, and 5) minimizing all patient activity. Decreasing the diastolic pressure below 50 would decrease coronary artery filling, which is definitely contraindicated in an ischemic setting. [Krupa, D. (Ed.). (1997). Flight nursing core curriculum. Park Ridge, IL: National Flight Nurses Association. (pp. 152).]

13. **b**
Patients in pulmonary edema do not need an increase in venous return and elevating the patients legs would increase venous return. A Semi-Fowlers position provides positioning to ease the work of breathing, increasing O2 percent increases the amount of oxygen available. Keeping patient pain free allows decrease in oxygen demand. [Krupa, D. (Ed.). (1997). Flight nursing core curriculum. Park Ridge, IL: National Flight Nurses Association. (pp. 158).]

14. **c**
Wedge pressure of 15 to 18 mm Hg indicates optimal left ventricular filling pressure. [Krupa, D. (Ed.). (1997). Flight nursing core curriculum. Park Ridge, IL: National Flight Nurses Association. (pp. 160).]

15. **b**
Appropriate treatment for a potential vascular injury includes frequent assessment and immobilization. Extremity fractures often require medication for pain relief and elevation to encourage venous return and to decrease edema. [Krupa, D. (Ed.). (1997). Flight nursing core curriculum. Park Ridge, IL: National Flight Nurses Association. (pp. 174).]

16. **c**
Initial management of a patient with a suspected cardiac tamponade begins with rapid intravenous fluid administration. This may transiently improve cardiac output while the medical team prepares for a pericardiocentesis. [Air and Surface Transport Nurses Association. (2002). Transport nurse advanced trauma course manual. Denver, CO. (pp. 125).]

17. **a**

Hypoxic hypoxia is defined as an oxygen deficiency in the body tissue sufficient to cause impaired function. These physiologic changes occur as the barometric pressure at altitude decreases, causing a reduction in the alveolar partial pressure of oxygen. [Holleran, R. (Ed.). (2003). Air and surface patient transport: Principals and practice. (3rd ed.). St. Louis: Mosby. (pp. 45).]

18. **d**

Defibrillation is the intervention of choice for ventricular fibrillation. Defibrillation in the air medical environment with current equipment in a medically equipped helicopter is safe providing the standard defibrillation precautions are observed. [Holleran, R. (Ed.). (2003). Air and surface patient transport: Principals and practice. (3rd ed.). St. Louis: Mosby. (pp. 375).]

19. **c**

The indications for emergency coronary artery bypass grafting include unstable angina unresponsive to medical therapy, evolving myocardial infarction with multi-vessel disease, and evolving MI when thrombolytic therapy or angioplasty are unsuccessful. [Holleran, R. (Ed.). (2003). Air and surface patient transport: Principals and practice. (3rd ed.). St. Louis: Mosby. (pp. 371).]

20. d

The mechanism of action for nitrates is to reduce cardiac oxygen demand by decreasing left-ventricular end-diastolic pressure (preload) and systemic vascular resistance (afterload). Nitrates also increase blood flow through the collateral coronary vessels via direct smooth muscle relaxation. [Cahill, M. (Ed) (1997). Nursing 97 drug handbook. Pennsylvania: Springhouse. (pp. 243-244).]

21. **c**

Management of patients with traumatic rupture of the diaphragm focuses on maintaining adequate oxygenation. This may require intubation and mechanical ventilation. [Air and Surface Transport Nurses Association. (2002). Transport nurse advanced trauma course manual. Denver, CO. (pp. 127-128).]

22. **a**

The most common cause of failure to pace is over sensing. Over sensing occurs when atrial activity is sensed during pacing, and a fast ventricular response will be noted. [Bojar, R. (1994). Manual of perioperative care in cardiac and thoracic surgery. Boston: Blackwell Scientific Publications. (pp. 6).]

23. **b**

The following are indications for pacing: symptomatic bradycardia unresponsive to drug therapies, overdriving tachydysrhythmias, and pulseless electrical activity due to electrolyte abnormalities. Although being prepared for pacing in a patient with asymptomatic bundle branch blocks, it is not an indication for emergency pacing. [Holleran, R. (Ed.). (2003). Air and surface patient transport: Principals and practice. (3rd ed.). St. Louis: Mosby. (pp. 375-376).]

24. **c**

Contraindications for thrombolytic therapy: history of intra-cranial bleeding or cerebrovascular accident, high risk of internal hemorrhage or active internal hemorrhage, uncontrolled hypertension, recent intracranial or intra-spinal surgery, intracranial tumor, arteriovenous malformation, and aneurysm and history of known bleeding diathesis. [Holleran, R. (Ed.). (2003). Air and surface patient transport: Principals and practice. (3rd ed.). St. Louis: Mosby. (pp. 369-370).]

25. **c**

The care of the patient with pre-existing intravascular hemodynamic monitoring lines in the air medical environment is attention to detail and careful handling to avoid potential complications. The flight nurse must label all lines clearly, secure all connections, place sterile caps over all exposed ports, and maintenance of a flush system during transport. [Holleran, R. (Ed.). (2003). Air and surface patient transport: Principals and practice. (3rd ed.). St. Louis: Mosby. (pp. 401-403).]

26. **b**

A normal central venous pressure (CVP) is 2-6 mmHg. The CVP in a patient with a right ventricular myocardial infarction (RVMI) will be consistent with RV failure demonstrating an elevated CVP and jugular venous distention. [Holleran, R. (Ed.). (2003). Air and surface patient transport: Principals and practice. (3rd ed.). St. Louis: Mosby. (pp. 401-403).]

Shock Transport

1. A 26-year-old man is hypotensive upon arrival at the emergency department following a motor vehicle crash. He is now to be transported to a Level 1 trauma facility. Assessment at this time reveals that he is obtunded, has a blood pressure of 102/84mm Hg, a heart rate of 108 beats/min, and shallow respirations of 24/mm. Initial interventions included administration of 3 L of lactated Ringer's solution and 1 unit of packed red blood cells. Arterial blood gas measurements on room air obtained 5 minutes ago revealed a pH of 7.24, a PaCO2 of 58 mm Hg, a PaO2 of 80 mm Hg, and a HCO3 of 19 mEq/L. The next step in intervention should be to:

   a. administer NaHCO3 1 mEq/kg.
   b. intubate and ventilate with 100% oxygen.
   c. administer oxygen via nasal cannula at 6 L/min.
   d. administer oxygen via non-rebreathing mask at 15 L/min.

2. A 23-year-old man is unconscious after a motor vehicle crash in which his car hit a tree and he was the unrestrained driver. Paramedics at the scene intubated the patient and established an IV with lactated Ringer's solution. Assessment reveals visible bruising over the sternum, equal breath sounds, and distended neck veins. He has a blood pressure of 80/40 mm Hg, a heart rate of 120 beats/mm, and assisted respirations of 16/mm. The next step in intervention would be to:

   a. perform pericardiocentesis.
   b. infuse dopamine at 20 mcg/kg/min.
   c. perform needle decompression on the left side of the chest.
   d. establish a second IV with lactated Ringer's solution, running both wide open.

3. Signs/symptoms of hypovolemic shock do not include:

   a. thirst.
   b. bradycardia.
   c. cool clammy skin.
   d. pallor and or cyanosis.

4. A beekeeper that has been stung multiple times is exhibiting signs of severe respiratory distress, along with urticaria, angioedema, and hypotension. Paramedics established an IV with normal saline solution. Appropriate intervention at this time is to:

   a. administer Diphenhydramine 50 mg IM.
   b. administer epinephrine 1:1000, 0.3 mg SQ.
   c. administer epinephrine 1:10000,0.5mg IV.
   d. initiate rapid sequence intubation and intubate the patient.

5. The normal blood volume for an adult is how many milliliters per kilogram (mL/kg)?

   a. 50 mL/Kg
   b. 60 mL/Kg
   c. 70 mL/Kg
   d. 80 mL/Kg

6. Current recommendations for use of a pneumatic antishock garment (PASG) include:

    a. stabilization of pelvic fractures.
    b. hypotension in congestive heart failure.
    c. respiratory distress due to chest injuries.
    d. symptomatic pericardial tamponade.

7. A multiple trauma patient has been in the intensive care unit of the local hospital for 24 hours. A central venous pressure line is in place with a current pressure of 18 cm of H20. All of the following are possible reasons for this pressure except:

    a. hypovolemia.
    b. pneumothorax.
    c. pulmonary edema.
    d. pericardial tamponade.

8. The average adult's hourly urinary output is approximately how many milliliters per kilogram (mL/kg)?

    a. 0.5 to l. mL/Kg
    b. 2 to 4. mL/Kg
    c. 5 to 7. mL/Kg
    d. 8 to 10. mL/Kg

9. A 27-year-old multiple trauma patient from the emergency department undergoes fluid resuscitation with 3 L of normal saline solution and 5 units of unwarmed packed red blood cells. He remains unconscious, intubated, and ventilated with 100% oxygen. He has received sedation and remains immobilized on a backboard. The flight nurse should remain concerned about:

    a. alkalosis due to blood administration.
    b. hypokalemia because of hemolyzed blood cells.
    c. hypothermia due to the use of unwarmed blood.
    d. decreased clotting times due to the use of banked blood.

10. A patient who sustains traumatic injuries to the head and pelvis is hypotensive and has been intubated. He is being adequately ventilated with 100% oxygen. Which of the following is a potential nursing diagnosis for this patient?

    a. Fluid volume deficit
    b. Ineffective airway clearance
    c. Impaired gas exchange due to ineffective ventilation
    d. Inadequate cardiac output related to myocardial dysfunction

11. The severity of perfusion failure may be best evaluated in a patient in septic shock by monitoring which of the following blood studies?

    a. CBC
    b. Serum potassium levels
    c. Serum sodium levels
    d. Serum lactate levels

12. A motorcyclist is thrown approximately 50 feet in a crash in which he was not wearing his helmet. He is unconscious, unresponsive, pale, and tachycardic with a blood pressure of 70 on palpation. He appears to have severe head and chest injuries. Paramedics intubated him and established an IV with normal saline solution at a keep open rate. The next step in intervention is to:

    a.   apply and inflate a PASG.
    b.   infuse dopamine at 5 mcg/kg/min.
    c.   elevate the head of the stretcher to decrease intracranial pressure.
    d.   increase flow rate to wide open and consider giving blood products.

13. A 55-year-old man with acute anterior wall myocardial infarction that is unresponsive to thrombolytic therapy is to be transported to a cardiac catheterization laboratory. The patient responds only to painful stimuli, and is pale with circumoral cyanosis and hypotension. He has a blood pressure of 70 on palpation, a heart rate of 118 beats/mm, and shallow respirations of 28/mm. Three IV lines are in place with dopamine and Lidocaine infusing. In addition, 2 L of normal saline solution have been administered. The next step in treatment is to:

    a.   apply and inflate a PASG.
    b.   administer packed red blood cells.
    c.   add Dobutamine and nitroglycerin to the drug therapy.
    d.   initiate rapid sequence intubation and hyperventilate with 100% oxygen.

14. Angiotensin II impacts the cardiovascular system by:

    a.   increasing both preload and afterload.
    b.   increasing preload and decreasing afterload.
    c.   decreasing both preload and afterload.
    d.   decreasing preload and increasing afterload.

15. Which of the following best describes the etiology of cardiogenic shock?

    a.   Renal failure
    b.   Pump failure
    c.   Inadequate volume
    d.   Inadequate vascular tone

16. Treatment of a patient in acute cardiogenic shock should include:

    a.   Aminophylline for bronchodilation.
    b.   heparin to prevent additional clots.
    c.   nitrates to decrease preload and afterload.
    d.   beta-blockers to decrease workload.

17. When monitoring a patient's hemodynamic status, which of the following parameters would best reflect left ventricular filling pressure?

    a.   Cardiac output
    b.   Mean arterial pressure
    c.   Central venous pressure
    d.   Pulmonary capillary wedge pressure

18.   The overall treatment goal for a patient in shock is:

    a.   to increase cardiac output.
    b.   to improve tissue perfusion.
    c.   to maintain the oxygen-carrying capacity of the hemoglobin.
    d.   All of the above.

19.   Which of the following phrases best defines anaphylactic shock?

    a.   Acute systemic reaction as a result of the release of chemical mediators
    b.   Loss of circulating volume
    c.   Massive stimulation of the sympathetic nervous system
    d.   Relative volume loss secondary to peripheral vasodilatation

20.   Treatment priorities for a patient in anaphylactic shock include administering:

    a.   Aminophylline for bronchodilation.
    b.   dopamine to support the blood pressure.
    c.   Dobutamine to increase cardiac output.
    d.   antihistamines and steroids to stop the inflammatory process.

Shock Transport Answers

1. b
The patient is obtunded, has shallow respirations and arterial blood gas results indicate respiratory acidosis. Intubating will control the airway and ventilation will correct the acidosis. A nasal cannula will not provide adequate oxygenation or ventilation in this shallow breathing patient. A non-rebreather mask does not provide ventilation to correct the acidosis. Sodium bicarbonate in not indicated as the respiratory acidosis can be corrected by ventilation. [Holleran, R. (Ed.). (2003). Air and surface patient transport: Principles and practice. (3rd ed.). St. Louis: Mosby. (pp. 213-214).]

2. a
Assessment reveals a probable pericardial tamponade. Pericardiocentesis is the definitive treatment. Needle decompression is inappropriate because there are equal lung sounds. A second IV line, may be appropriate after relief of the tamponade. Dopamine is not indicated until the tamponade has been relieved and fluids replaced. [Holleran, R. (Ed.). (2003). Air and surface patient transport: Principles and practice. (3rd ed.). St. Louis: Mosby. (pp. 284-285).]

3. b
Baroreceptors decrease vagal response and increase sympathetic tone resulting in tachycardia. Thirst is the result of decreased fluids. Cool, clammy skin is a result of increased peripheral vascular resistance resulting in improved flow to vital organs. Pallor and or cyanosis are the result of decreased tissue oxygenation. [Krupa, D. (Ed.). (1997). Flight nursing core curriculum. Park Ridge, IL: National Flight Nurses Association. (pp. 190-191).]

4. c
The patient is exhibiting severe anaphylaxis, is hypotensive and requires IV epinephrine. Diphenhydramine is indicated in less severe reactions and may be appropriate after administration of the IV epinephrine. Rapid sequence induction (RSI) and intubation may be necessary, however, the epinephrine may eliminate the need to intubate the patient. SQ epinephrine is indicated for less severe reactions. [Holleran, R. (Ed.). (2003). Air and surface patient transport: Principles and practice. (3rd ed.). St. Louis: Mosby. (pp. 220-221).]

5. c
The normal blood volume is 70 ml/Kg. [Holleran, R. (Ed.). (2003). Air and surface patient transport: Principles and practice. (3rd ed.). St. Louis: Mosby. (pp. 217).]

6. a
The pneumatic antishock garment (PASG) provides stabilization for the fractured pelvis. No significant improvement in outcome has been demonstrated when compared to fluids and rapid transport. Inflation of the PASG may result in increased bleeding from areas not covered by the garment, including supradiaphragmatic injuries. Pericardial tamponade requires pericardiocentesis. Pulmonary edema is an absolute contraindication. [Holleran, R. (Ed.). (2003). Air and surface patient transport: Principles and practice. (3rd ed.). St. Louis: Mosby. (pp. 215).]

7. a
The normal CVP is 4-10 cm H2O. Readings above 10 cm H2O are due to excessive preload (fluid overload), elevated pulmonary system pressure (pneumothorax and pulmonary edema), or the inability of the left ventricle to pump blood (pericardial tamponade). [Holleran, R. (Ed.). (2003). Air and surface patient transport: Principles and practice. (3rd ed.). St. Louis: Mosby. (pp. 401-402).]

8. a

A urinary output of less than 35 ml/hr indicates shock. The average adult weighs 70 kg, therefore 0.5ml/kg is the average hourly output. The remaining choices indicate diuresis. [Holleran, R. (Ed.). (2003). Air and surface patient transport: Principles and practice. (3rd ed.). St. Louis: Mosby. (pp. 198).]

9. c

The unwarmed blood may cause hypothermia, which results in shift of the oxyhemoglobin curve to the left. The opposites of A, B, C, are possible with blood administration. [Holleran, R. (Ed.). (2003). Air and surface patient transport: Principles and practice. (3rd ed.). St. Louis: Mosby. (pp. 417).]

10. a

The patient is hypotensive, possibly due to blood loss from the pelvic fracture. Ineffective airway clearance and impaired gas exchange due to ineffective ventilation are related to airway and respiratory involvement, which has been corrected with the ET. Inadequate cardiac output related to myocardial dysfunction is related to cardiac injury, which is not indicated in this patient. [Cardona, V. (Ed.). (1994). Trauma nursing. (2nd ed.). Philadelphia: W. B. Saunders. (pp. 156-58).]

11. d

While CBC, Na levels and K levels provide important information on the patient's condition, lactate levels alone are a particularly good indicator of tissue perfusion. [Cardona, V. (Ed.). (1994). Trauma nursing. (2nd ed.). Philadelphia: W. B. Saunders. (pp. 174-75).]

12. d

Despite the head injury, the patient is tachycardic and hypotensive so the flight nurse should suspect another source of bleeding. In addition, it is necessary to maintain cerebral perfusion pressure (CPP) to minimize further head injury. The PASG will increase blood pressure, however fluid resuscitation is indicated. Dopamine is not indicated until fluids have been administered. Elevating the head of the cot may reduce intracranial pressure, however the patient is hypotensive and the mean arterial pressure must be restored to assure cerebral perfusion pressure. [Holleran, R. (Ed.). (2003). Air and surface patient transport: Principles and practice. (3rd ed.). St. Louis: Mosby. (pp. 268-269).]

13. d

The patient is not ventilating well and his airway must be secured to assure adequate oxygenation. Administration of PRBCs may be necessary but the airway and breathing take precedence. PASG is not indicated. Dobutamine and nitroglycerin may be appropriate for pulmonary edema, but again the airway takes precedence. [Holleran, R. (Ed.). (2003). Air and surface patient transport: Principles and practice. (3rd ed.). St. Louis: Mosby. (pp. 376-378).]

14. a

Angiotensin II is a potent vasoconstrictor. Angiotensin II also causes the release of aldosterone, which results in reabsorption of sodium and water in the renal tubules. With the reabsorption of volume, preload is increased and with vasoconstriction, afterload is increased. [Emergency Nurses Association. (2000). Trauma nursing core course instructor manual. (5th ed.). Chicago, IL. (pp. 83-84).]

15. b

The etiology of cardiogenic shock is the failure of the "pump", i.e. the heart. [Krupa, D. (Ed.). (1997). Flight nursing core curriculum. Park Ridge, IL: National Flight Nurses Association. (pp. 198).]

16. **c**

Treatment of cardiogenic shock is to administer to decrease preload by decreasing the volume return to an already compromised pump. Nitrates also decrease afterload, the pressure the heart has to pump against, by causing peripheral vasodilatation. Dopamine will help maintain arterial pressure and renal perfusion at lower doses allowing nitrate use as well as maintaining glomerular filtration to decrease preload. Dobutamine increases contractility and thus increases cardiac output to assist 'in maintaining peripheral perfusion. [Krupa, D. (Ed.). (1997). Flight nursing core curriculum. Park Ridge, IL: National Flight Nurses Association. (pp. 200).]

17. **d**

Pulmonary capillary wedge pressure reflects left ventricular filling pressure and is clinically useful as a marker for fluid administration or restriction. [Holleran, R. (Ed.). (2003). Air and surface patient transport: Principles and practice. (3rd ed.). St. Louis: Mosby. (pp. 401-402).]

18. **b**

The overall goal in the management of shock is to improve tissue perfusion. Shock is a manifestation of cellular insufficiency. The common denominator of all shock states is the amount of oxygen consumed by the cells. [Holleran, R. (Ed.). (2003). Air and surface patient transport: Principles and practice. (3rd ed.). St. Louis: Mosby. (pp. 207-215).]

19. **a**

Anaphylaxis is an acute systemic allergic reaction as a result of the release of chemical mediators after an antigen-antibody reaction. [Holleran, R. (Ed.). (2003). Air and surface patient transport: Principles and practice. (3rd ed.). St. Louis: Mosby. (pp. 220).]

20. **d**

Airway is the number one priority. Epinephrine is given to dilate bronchial smooth muscles causing vasoconstriction and decreasing permeability. An antihistamine will compete with the binding sites and administration of steroids is also needed. [Holleran, R. (Ed.). (2003). Air and surface patient transport: Principles and practice. (3rd ed.). St. Louis: Mosby. (pp. 221).]

Musculoskeletal and Multiple Trauma Transport

1. Long-term complications associated with musculoskeletal injuries include all of the following except:

    a. infection.
    b. thrombophlebitis.
    c. delayed bone healing.
    d. adult respiratory distress syndrome.

2. Which of the following statements about fractures of the extremities is false?

    a. The risk of infection associated with both open and closed fractures is the same.
    b. An open wound near a fracture site should be considered an open fracture until proven otherwise.
    c. Assessment and treatment of extremity fractures should focus on potential neurovascular compromise.
    d. Management of hand or foot fractures should focus on maximizing return of function and preventing permanent disability.

3. Spontaneous recurrence is commonly associated with which of the following types of dislocations?

    a. Anterior hip
    b. Posterior hip
    c. Anterior shoulder
    d. Posterior shoulder

4. At room temperature, an amputated part is viable for possible reimplantation for approximately how many hours?

    a. 1 to 3 hours
    b. 4 to 6 hours
    c. 7 to 9 hours
    d. More than 10 hours

5. In packaging a patient with a traumatic amputation, the flight crew should do all of the following in regards to the amputated part except:

    a. located and transported with the patient.
    b. Keep part cool by placing it in a dry sterile dressing and placing part in bag. This bag should then be placed in a second sealed bag, filled with ice.
    c. wrapped in a dry, sterile gauze dressing or towel, and placed in a sealed bag.
    d. maintained as it was found, with any clothing or shoes left in place, as removing these may cause further damage.

6. A late developing sign of fat embolism syndrome is:

    a. peripheral cyanosis.
    b. mental status changes.
    c. petechial hemorrhages.
    d. increased work of breathing.

7. The mnemonic "P.M.S." used for assessing a patient with a musculoskeletal injury refers to which of the following?

    a. Pain, movement, sensation
    b. Pain, range of motion, sensation/pain
    c. Palpable pulses, movement, stabilization
    d. Palpable pulses, movement, sensation/pain

8. Which of the following nursing diagnoses is associated with an expected event in all patients with fractures?

    a. Impaired gas exchange related to prolonged immobility
    b. Activity intolerance related to anemia and/or injured state
    c. Pain subsequent to the injury and the inflammatory process
    d. Hypothermia related to prolonged exposure to cool ambient temperatures

9. Death resulting from crush syndrome is most commonly due to:

    a. infection.
    b. renal failure.
    c. metabolic acidosis.
    d. adult respiratory distress syndrome.

10. Complications of crush injuries and crush syndrome include which of the following:

    a. renal failure, compartment syndrome, adult respiratory distress syndrome.
    b. disseminating intravascular coagulopathy, adult respiratory distress syndrome.
    c. renal failure, compartment syndrome, disseminating intravascular coagulopathy.
    d. renal failure, adult respiratory distress syndrome, disseminating intravascular coagulopathy.

11. A potential nursing diagnosis related to a closed mid shaft femur fracture is:

    a. fluid volume deficit related to hemorrhage.
    b. risk of infection.
    c. impaired gas exchange.
    d. ineffective thermoregulation.

12. The most common musculoskeletal dislocation is the:

    a. shoulder.
    b. elbow.
    c. knee.
    d. hip.

13. Signs and symptoms of a right anterior hip dislocation include all the following except:

    a. wide abduction.
    b. external rotation.
    c. right leg shortened.
    d. extremity flexed.

14. Compartment syndrome:

   a. develops rapidly, often in the first 30 minutes post injury.
   b. pain that increases with active muscle stretching.
   c. compartment pressure of 20-25 mmHg indicates need for emergent fasciotomy.
   d. diminishing distal pulses and delayed capillary refill are late signs.

15. A 26-year-old construction worker sustains blunt trauma after a fall of three stories. Initial assessment reveals a blood pressure of 76/40 mm Hg, a heart rate of 132 beats/min, and respirations of 36/min. There are no breath sounds on the right side, a shift of the point of maximum intensity to the left, flat neck veins, and hemoptysis. These signs and symptoms suggest:

   a. an aortic rupture.
   b. cardiac tamponade.
   c. massive hemothorax.
   d. tension pneumothorax.

16. Initial assessment and interventions for a patient who sustained multiple trauma as a result of a motor vehicle crash would not include which of the following steps?

   a. Initiating two large-bore IVs
   b. Conducting a head-to-toe examination
   c. Placing an occlusive dressing on a sucking chest wound
   d. Maintaining the airway while immobilizing the cervical spine

17. Interventions for an adult who sustained multiple trauma are expected to result in:

   a. urine output of 15 ml/hour.
   b. secure and patent airway.
   c. decreasing level of consciousness.
   d. blood pressure of 60/palp, heart rate of 160 beats/min, and respirations of 24/min.

18. A patient to be air transported has an open chest wound on the right side and an open fracture of the left femur. Treatment by the ground paramedic included intubation, establishing IV access, and placing an occlusive dressing over his right chest wall. After lift-off, the patient suddenly becomes cyanotic and more tachycardic. The flight crew should first:

   a. extubate and re-intubate.
   b. remove the chest wall dressing.
   c. open the IV fluids to wide open.
   d. perform needle decompression on the right side of the chest.

19. A priority nursing diagnosis in the multiple trauma patient would include all of the following except:

   a. actual fluid volume deficit.
   b. ineffective airway clearance.
   c. impaired verbal communication.
   d. potential for impaired gas exchange.

Musculoskeletal and Multiple Trauma Transport Answers

1. **b**
Thrombophlebitis is a complication that usually presents during the patients initial hospitalization and is considered an early complication, not a long-term complication. It is usually caused by immobility and venous stasis. [Krupa, D. (Ed.). (1997). Flight nursing core curriculum. Park Ridge, IL: National Flight Nurses Association. (p. 227).]

2. **a**
Open fractures have a much higher associated infection rate than closed fractures. Open wounds near the fracture site should be treated as an open fracture until proven otherwise. [Krupa, D. (Ed.). (1997). Flight nursing core curriculum. Park Ridge, IL: National Flight Nurses Association. (p. 230).]

3. **c**
Anterior shoulder dislocations are associated with younger patients and may spontaneously recur with no associated trauma. [Krupa, D. (Ed.). (1997). Flight nursing core curriculum. Park Ridge, IL: National Flight Nurses Association. (p. 235).]

4. **b**
Replantation of an amputated extremity should be done as soon as possible, but an extremity preserved at room temperature may still be considered viable for 4-6 hours. Proper cooling of the amputated part will increase the viability of the amputated part and replantation may be considered for up to 18 hours after injury. [Krupa, D. (Ed.). (1997). Flight nursing core curriculum. Park Ridge, IL: National Flight Nurses Association. (p. 237).]

5. **d**
All clothing, shoes and major debris should be removed from the amputated part prior to cooling. Constriction or the possible tourniquet effect of tight clothing may cause further tissue damage and may decrease the chance of successful replantation of the amputated part. [Holleran, R. (Ed.). (2003). Air and surface patient transport: Principles and practice. (3rd ed.). St. Louis: Mosby. Air and Surface Transport Nurses Association. (p. 318)].

6. **c**
Petechial hemorrhages are considered to be a sign of developing fat embolus syndrome and will usually be seen late in the course of this medical complication. [Krupa, D. (Ed.). (1997). Flight nursing core curriculum. Park Ridge, IL: National Flight Nurses Association. (p. 242).]

7. **d**
The use of the mnemonic "P.M.S." stands for the following: Palpable Pulses, Movement, Sensation/Pain. It is easily remembered and can be used to assess initial and subsequent changes in the neurovascular status of a fractured extremity. [Krupa, D. (Ed.). (1997). Flight nursing core curriculum. Park Ridge, IL: National Flight Nurses Association. (p. 231).]

8. **c**
Pain subsequent to the injury and the inflammatory process would be an expected event in all patients with fractures. Not all patients would experience respiratory complications, activity intolerance or hypothermia. [Krupa, D. (Ed.). (1997). Flight nursing core curriculum. Park Ridge, IL: National Flight Nurses Association. (p. 229).]

9. **b**
Death associated with crush injuries and resulting crush syndromes are most frequently caused by renal failure. Renal failure occurs because skeletal muscle injury can lead to myoglobinuria and renal failure secondary to acute tubular necrosis. [Krupa, D. (Ed.). (1997). Flight nursing core curriculum. Park Ridge, IL: National Flight Nurses Association. (p. 238).]

10. **c**
Complication of crush syndrome includes renal failure, compartment syndrome and disseminating intravascular coagulopathy. Adult respiratory distress syndrome is more commonly associated with fat embolism. [Krupa, D. (Ed.). (1997). Flight nursing core curriculum. Park Ridge, IL: National Flight Nurses Association. (pp. 238-239).]

11. **a**
A mid-shaft femur fracture injury has a high risk for vascular injury, so a fluid volume deficit is likely. As a closed fracture, the risk of infection from the injury is not likely. [Krupa, D. (Ed.). (1997). Flight nursing core curriculum. Park Ridge, IL: National Flight Nurses Association. (p. 232).]

12. **d**
Posterior hip dislocation is the most common of dislocations. [Krupa, D. (Ed.). (1997). Flight nursing core curriculum. Park Ridge, IL: National Flight Nurses Association. (p. 235).]

13. **d**
Anterior hip dislocations occur with deformity and wide abduction, external rotation of the extremity, and the affected leg is noticeably shortened. A posterior hip dislocation has the extremity flexed and adducted. [Krupa, D. (Ed.). (1997). Flight nursing core curriculum. Park Ridge, IL: National Flight Nurses Association. (p. 235).]

14. **d**
Diminishing distal pulses and delayed capillary refill are late signs and indicate already high compartment pressures. Compartment syndrome develops over hours after the injury occurs. Pain increases with passive muscle stretching. Pressure > 30mmHg suggest a need for fasciotomy. [Krupa, D. (Ed.). (1997). Flight nursing core curriculum. Park Ridge, IL: National Flight Nurses Association. (p. 240).]

15. **c**
You would suspect the patient to have a massive hemothorax. A patient with a massive hemothorax that results in hypovolemia may exhibit flat neck veins. With a tension pneumothorax or cardiac tamponade, you would expect to see distended neck veins. A tension pneumothorax, cardiac tamponade or aortic rupture does not usually result in hemoptysis. [Krupa, D. (Ed.). (1997). Flight nursing core curriculum. Park Ridge, IL: National Flight Nurses Association. (pp. 126,129, 215).]

16. **b**
Primary survey assessment and interventions focus on airway, breathing, and circulation and treatment of life-threatening injuries found. The head to toe examination is a part of the secondary survey. [Krupa, D. (Ed.). (1997). Flight nursing core curriculum. Park Ridge, IL: National Flight Nurses Association. (pp. 213, 214, 216).]

17. **b**
A secure and patent airway is an expected outcome with an adult multiple trauma patient. Urine output of greater than 30cc/hr is expected, as are vital signs within normal limits and an increasing level of consciousness. [Krupa, D. (Ed.). (1997). Flight nursing core curriculum. Park Ridge, IL: National Flight Nurses Association, (p. 224).]

18. **b**
If a tension pneumothorax develops with an occlusive dressing over an open pneumothorax, the chest dressing should be removed (or lifted) to decompress the tension pneumothorax. [Krupa, D. (Ed.). (I 997). Flight nursing core curriculum. Park Ridge, IL: National Flight Nurses Association. (p. 131).]

---

19. **c**

Ineffective airway clearance, potential for impaired gas exchange and actual fluid volume deficit all relate to the primary survey. Impaired verbal communication is not a priority nursing diagnosis in the multiple trauma patient. [Krupa, D. (Ed.). (1997). Flight nursing core curriculum. Park Ridge, IL: National Flight Nurses Association. (pp. 213-216, 222).]

Surface Trauma and Toxicological Emergencies Transport

1. What is the appropriate nursing diagnosis for a patient with a hymenoptera sting?

    a. High risk for self-injury
    b. Altered tissue integrity
    c. Ineffective thermal regulation
    d. Fluid volume deficit due to hemorrhage

2. All of the following regarding preservation of a violent crime scene are true except:

    a. avoid removing gunpowder from the skin.
    b. do not touch or move weapons or other objects.
    c. avoid cutting clothes through stab or bullet holes.
    d. place items in plastic bags to be given to law enforcement.

3. Local anesthetics containing epinephrine should be avoided in patients taking beta-blockers because:

    a. hypotension may result when the epinephrine opposes beta activity.
    b. hypotension may result when the epinephrine triggers a paradoxical beta response.
    c. a hypertensive crisis may be triggered when the epinephrine enhances beta activity.
    d. a hypertensive crisis may be triggered when alpha- stimulating activity is   unopposed by beta activity.

4. Management of a contaminated wound with avulsed tissue includes all of the following except:

    a. administering a local anesthetic, such as 1 % Lidocaine with epinephrine.
    b. irrigating the wound to remove gross contamination.
    c. dressing the wound with a petroleum type dressing.
    d. applying a splint to immobilize the affected part.

5. Treatment and antivenin dosage is determined based on the envenomation severity scale, which includes all of the following criteria except the:

    a. skin temperature at the site of envenomation.
    b. extent of edema at the site of envenomation.
    c. presence of systemic symptoms.
    d. severity of pain.

6. A patient with a snakebite injury should be placed in which of the following positions?

    a. Supine
    b. Left lateral
    c. Semi-Fowler
    d. High Fowler's

7. Of the following conditions, which does not delay or adversely affect wound healing?

   a. Diabetes mellitus
   b. Blood dyscrasias
   c. Coronary artery disease
   d. Peripheral vascular disease

8. Important scene information regarding gunshot wounds includes all the following except:

   a. caliber of the weapon.
   b. type and caliber of the bullet.
   c. time of injury.
   d. name of person who shot the patient.

9. Puncture wounds are at high risk for:

   a. significant external hemorrhage.
   b. infection.
   c. severe pain.
   d. none of the above.

10. A degloving injury is a form of:

   a. laceration.
   b. abrasion.
   c. avulsion.
   d. puncture.

11. Infections that can occur from a human bite include all the following except:

   a. hepatitis.
   b. actinomycosis.
   c. tetanus.
   d. rabies.

12. The most dangerous location for snakebites is?

   a. Trunk
   b. Upper extremity
   c. Lower extremity
   d. All are equally dangerous

13. Brown recluse spider bites have the following reaction:

   a. necrotic lesion.
   b. presence of tiny fang marks.
   c. pain in the first 15 minutes.
   d. abdominal rigidity.

14. Individuals in which of the following occupations are at a high risk for cyanide poisoning?

    a. Gardening
    b. Spray painting
    c. Jewelry making
    d. Stained glass artistry

15. Which of the following are signs and symptoms of Salicylate poisoning?

    a. Tinnitus, thirst, headache, and elevated temperature
    b. Dizziness, decreased respirations, nystagmus, and respiratory arrest
    c. Nausea, right upper quadrant pain and tenderness, pallor, and diaphoresis
    d. Fatigue, visual disturbances, hallucinations, and cardiac rhythm disturbances

16. Alcohol poisoning patients are frequently found to be hypothermic due to what effect of alcohol on the body?

    a. A decrease in metabolic rate
    b. Enhanced cutaneous blood flow
    c. A decreased level of consciousness
    d. Depression of the temperature regulator of the brainstem

17. Activated charcoal can prevent absorption of a poison that was ingested orally by:

    a. inducing emesis.
    b. decreasing gastric acidity.
    c. increasing gastrointestinal motility.
    d. absorbing poisons in the gastrointestinal tract.

18. A patient who ingests a poison should be placed in what position in order to delay absorption?

    a. Supine
    b. Trendlenburg
    c. Left lateral decubitus
    d. Right lateral decubitus

19. What is the antidote for an opiate poisoning?

    a. Naloxone
    b. Flumazenil
    c. Pralidoxime
    d. Methylene blue

20. Acetaminophen poisoning has been prevented if the result of which of the following tests is within normal limits?

    a. White blood cell count
    b. Liver function tests
    c. PaO2 and PaCO2 levels
    d. Cardiac enzyme levels

21.   Intervention for a patient with alcohol poisoning has been successful if results of which of the following tests are within normal limits?

       a.   Cardiac enzyme levels
       b.   Blood glucose levels
       c.   Creatinine blood levels
       d.   White blood cell counts

22.   Prevention of further toxicity in an organophosphate poisoning is evidenced by:

       a.   clear lungs.
       b.   decreased heart rate.
       c.   increased respirations.
       d.   increased moisture of the mucous membranes.

Surface Trauma and Toxicological Emergencies Transport Answers

1. b
Hymenoptera stings (bees, wasps, hornets, and ants) cause an interruption in the skin, thereby creating an alteration in tissue integrity. [Krupa, D. (Ed.). (1997). Flight nursing core curriculum. Park Ridge, IL: National Flight Nurses Association. (pp. 274).]

2. d
Items should be placed in paper bags not plastic. Condensation can occur with plastic bags thereby destroying evidence. The area of dispersal of gun powder can give clues to the distance between the patient and the weapon that was fired, if removal of the gun powder is necessary for patient care, careful documentation of the wound and presence of gunpowder need to be done. The location of evidence at a scene and any damage to clothing also provides clues to law enforcement, unless necessary for patient care leaving evidence where it lies and cutting clothing to allow the damage to remain intact can help the law enforcement agency determine important information regarding the crime. [Krupa, D. (Ed.). (1997). Flight nursing core curriculum. Park Ridge, IL: National Flight Nurses Association. (pp. 250).]

3. d
A hypertensive crisis may be triggered when alpha-stimulating activity is unopposed by beta activity. the alpha properties of epinephrine are normally opposed by beta responses, thereby preventing hypertension. Patients taking beta-blocking medication do not have this normal response, and may develop hypertensive crisis. [Krupa, D. (Ed.). (1997). Flight nursing core curriculum. Park Ridge, IL: National Flight Nurses Association. (pp. 256).]

4. a
Administering a local anesthetic, such as 1% Lidocaine with epinephrine. Anesthetics containing epinephrine should be avoided in contaminated wounds. Epinephrine induces local ischemia that interferes with the mobilization of immune mechanisms needed to prevent wound infection. While traumatic wounds should not distract attention away from more serious the potential for wound infection must be considered in less critical instances and when wounds have been contaminated for an extended period of time. Immobilization of the affected part prevents further trauma during transport. A non-adhering dressing prevents further tissue trauma and debriding of tissue when dressings are removed. Dry gauze may be applied afterward to maintain wound asepsis. [Krupa, D. (Ed.). (1997). Flight nursing core curriculum. Park Ridge, IL: National Flight Nurses Association. (pp. 259).]

5. a
Skin temperature at wound sites can indicate co-morbidity such as local infection, ischemia, etc., but is not a factor in envenomation severity grading. Systemic neurologic and hematologic symptoms indicate a severe response. Rapid and extensive swelling at the envenomation site is indicative of a severe response. Girth of the involved extremity should be measured every 15 minutes. While most snakebites are painful severe responses cause wide spread pain, extending well beyond the immediate envenomation site. [Krupa, D. (Ed.). (1997). Flight nursing core curriculum. Park Ridge, IL: National Flight Nurses Association. (pp. 267).]

6. a
When transporting snakebite victims, the preferred position is supine to encourage rest. Rest will decrease the metabolism, spread, and absorption of venom. [Krupa, D. (Ed.). (1997). Flight nursing core curriculum. Park Ridge, IL: National Flight Nurses Association. (pp. 267).]

7. **c**

Disorder/diseases affecting peripheral circulation and clotting mechanisms can delay or adversely affect wound healing. Severe heart disease can produce low cardiac output states, but typically does not adversely affect wound healing. Poor peripheral circulation can result in delayed wound healing and increased incidence of infection. Blood dyscrasias include several different syndromes such as hemophilia and Von Wildebrand's. These disorders prolong clotting time and can lead to increased bleeding with hemorrhage and anemia. In addition to adversely affecting peripheral circulation, diabetes mellitus increases a patient's risk for wound infection due to hyperglycemia. [Krupa, D. (Ed.). (1997). Flight nursing core curriculum. Park Ridge, IL: National Flight Nurses Association. (pp. 247).]

8. **d**

The caliber of the weapon and bullet can assist the trauma team in estimating potential internal injuries. Time since injury allows for estimation of the urgency for care. Who shot the person is not of any use to the car of the patient initially, however, where the person is may be a significant scene safety issue. [Krupa, D. (Ed.). (1997). Flight nursing core curriculum. Park Ridge, IL: National Flight Nurses Association. (pp. 251).]

9. **b**

Characteristically puncture wounds bleed minimally and tend to seal off creating a high potential for infection. [Krupa, D. (Ed.). (1997). Flight nursing core curriculum. Park Ridge, IL: National Flight Nurses Association. (pp. 253).]

10. c

A degloving a severe type of avulsion where full thickness skin appears to be "peeled away" from underlying tissue. [Krupa, D. (Ed.). (1997). Flight nursing core curriculum. Park Ridge, IL: National Flight Nurses Association. (pp. 254).]

11. **d**

Rabies is a risk with bites from animals, not humans. [Krupa, D. (Ed.). (1997). Flight nursing core curriculum. Park Ridge, IL: National Flight Nurses Association. (pp. 262).]

12. **a**

Snakebites on the head and trunk are 2 – 3 times more dangerous than on the extremities. Those on the upper extremities are more serious than those on the lower extremities. [Krupa, D. (Ed.). (1997). Flight nursing core curriculum. Park Ridge, IL: National Flight Nurses Association. (pp. 266).]

13. **a**

A brown recluse spider bite may have no lesion or a necrotic on as large as 30 cm. The other three items are characteristic of a black widow spider bite. [Krupa, D. (Ed.). (1997). Flight nursing core curriculum. Park Ridge, IL: National Flight Nurses Association. (pp. 270-271).]

14. **c**

Jewelers are exposed to cyanide during jewelry making. Gardeners have an exposure risk to pesticides, while spray painters are exposed to carbon monoxide, and stained glass artists use lead. [Krupa, D. (Ed.). (1997). Flight nursing core curriculum. Park Ridge, IL: National Flight Nurses Association. (pp. 589).]

15. **a**

Salicylate poisoning causes tinnitus, thirst, and headache, along with elevated temperature due to increased metabolic rate. Nausea, right upper quadrant pain and tenderness, pallor and diaphoresis are signs and symptoms of acetaminophen toxicity. Dizziness, decreased respiration, nystagmus, and respiratory arrest are signs and symptoms of sedative/hypnotic toxicity. Fatigue, visual disturbances, hallucinations, and cardiac rhythm disturbances are signs and symptoms of cardiac glycoside poisoning. [Krupa, D. (Ed.). (1997). Flight nursing core curriculum. Park Ridge, IL: National Flight Nurses Association. (pp. 594).]

16. **b**

Alcohol enhances blood flow to the skin, causing an increase in heat loss, which contributes to hypothermia. [Krupa, D. (Ed.). (1997). Flight nursing core curriculum. Park Ridge, IL: National Flight Nurses Association. (pp. 619).]

17. **d**

Charcoal absorbs many drugs and poisons in the gastrointestinal tract, preventing absorption into the bloodstream. Syrup of ipecac induces emesis, but its use is controversial. Charcoal does not affect gastric acidity or gastrointestinal motility. Charcoal is often combined with magnesium citrate or sulfates, both of which are cathartics and act to increase gastrointestinal motility. [Krupa, D. (Ed.). (1997). Flight nursing core curriculum. Park Ridge, IL: National Flight Nurses Association. (pp. 592-593).]

18. **c**

The left lateral decubitus position situates the pyloric valve so gastric emptying into the small intestine is more difficult, which delays absorption and allows more time for gastric decontamination. [Krupa, D. (Ed.). (1997). Flight nursing core curriculum. Park Ridge, IL: National Flight Nurses Association. (pp. 591).]

19. **a**

Naloxone is the antidote for opiates. Flumazenil is the antidote for pure benzodiazepine overdoses, while pralidoxime is the antidote for organophosphate poisoning, and Methylene blue is used with nitrite and nitrate poisonings. [Krupa, D. (Ed.). (1997). Flight nursing core curriculum. Park Ridge, IL: National Flight Nurses Association. (pp. 591).]

20. **b**

Acetaminophen is toxic to the liver, therefore monitoring liver enzymes as well as acetaminophen levels can assist in the evaluation of the patient. Respiratory function, blood counts, and cardiac enzymes are not affected. [Krupa, D. (Ed.). (1997). Flight nursing core curriculum. Park Ridge, IL: National Flight Nurses Association. (pp. 596-597).]

21. **b**

Alcohol poisoning creates a hypoglycemia, so the blood glucose level needs to be monitored during treatment. Other lab values that may be abnormal are liver function tests, renal function tests, blood gases, and clotting values. [Krupa, D. (Ed.). (1997). Flight nursing core curriculum. Park Ridge, IL: National Flight Nurses Association. (pp. 621-622).]

22. **a**

Muscarinic affects in organophosphate poisonings cause a dramatic increase in bronchial secretions, so the lung being clear of secretions is an indication of an appropriate response to treatment. The other answers are incorrect because the response of the muscarinic receptors to organophosphate poisoning cause bradycardia, increase in salivation, and the bronchospasm and secretions cause an increase in respiratory rate. [Krupa, D. (Ed.). (1997). Flight nursing core curriculum. Park Ridge, IL: National Flight Nurses Association. (pp. 615-616).]

Maxillofacial, Anterior Neck, and Eye Transport

1.   The facial bones most commonly fractured in blunt trauma are:

     a.   nose and zygoma.
     b.   nose and mandible.
     c.   maxilla and zygoma.
     d.   maxilla and mandible.

2.   Appropriate nursing interventions for globe extrusion include all of the following responses except:

     a.   attempting to replace the globe.
     b.   maintaining cervical spine precautions, as necessary.
     c.   applying bilateral dressings to decrease eye movement.
     d.   applying a moist saline dressing to stabilize the globe in place.

3.   Which of the following eye injuries requires the most emergent nursing intervention?

     a.   Chemical burn
     b.   Globe rupture
     c.   Retinal detachment
     d.   Penetrating trauma

4.   Nursing intervention for a patient with anterior neck trauma may include:

     a.   cricothyrotomy.
     b.   cervical spine immobilization.
     c.   use of a smaller size endotracheal tube, if intubation becomes necessary.
     d.   all of the above steps are appropriate.

5.   The highest priority nursing diagnosis for a patient with anterior neck trauma would be the potential for:

     a.   infection.
     b.   fluid volume deficit.
     c.   impaired gas exchange.
     d.   ineffective airway clearance.

6.   Which of the following is considered an early sign of an air embolus in a trauma patient who sustains a vascular injury to the anterior neck?

     a.   Apnea
     b.   Bradycardia
     c.   Hypertension
     d.   Decreased level of consciousness

7.   Effective airway clearance in a patient with anterior neck trauma may be indicated by the presence of:

     a.   an open airway.
     b.   bilateral equal breath sounds.
     c.   bilateral equal chest excursions with ventilations.
     d.   all of the above.

8. Which of the following findings is considered a sign of a fractured larynx?

   a. Neck pain
   b. Dysphagia
   c. Subcutaneous emphysema
   d. Enhanced cricothyroid prominence

9. Which of the following nursing diagnoses is specific to a patient with an esophageal injury?

   a. Potential for pain
   b. Potential for infection
   c. Potential for ineffective airway clearance
   d. All of the above

10. Early palpation and inspection of the neck in a patient believed to have a fractured larynx may reveal:

   a. a tracheal shift.
   b. neck vein distention.
   c. cricothyroid prominence.
   d. subcutaneous emphysema.

11. Cranial nerve III is associated with:

   a. pupil constriction.
   b. visual fields.
   c. inferior and medial eye movement.
   d. eyelid closing.

12. A Le Fort II fracture is also known as a:

   a. transverse fracture.
   b. pyramidal fracture.
   c. craniofacial dysjunction.
   d. alveolar arch fracture.

13. A hyphema is defined as:

   a. blood in the anterior chamber.
   b. conjunctival injection.
   c. loss of acute angle closure.
   d. retinal detachment.

Maxillofacial, Anterior Neck, and Eye Transport Answers

1. **b**
The most frequently fractured facial bone is the nose due to its prominence . The second most frequently fractured facial bone is the mandible. [Krupa, D. (Ed.). (1997). Flight nursing core curriculum. Park Ridge, IL: National Flight Nurses Association. (pp. 290, 293).]

2. **a**
Interventions for an extrusion of the globe include stabilizing the globe, and applying moist saline dressings without attempting to replace the globe. Lightly patch both eyes to minimize movement and decrease further damage. Any eye injury needs to be assessed for the possibility of cervical spinal injury, possibly requiring spinal immobilization. [Krupa, D. (Ed.). (1997). Flight nursing core curriculum. Park Ridge, IL: National Flight Nurses Association. (pp. 330).]

3. **a**
Chemical burns are considered the most emergent ocular injury as seconds can count in prevention of long-term injury. [Krupa, D. (Ed.). (1997). Flight nursing core curriculum. Park Ridge, IL: National Flight Nurses Association. (pp. 322).]

4. d
Interventions for any patient with anterior neck trauma will include cervical spine immobilization. Any mechanism causing anterior neck trauma creates a high index of suspicion for injury to the cervical spine. If intubation is required a smaller size ETT should be used to ensure passage without causing irritation and creating more swelling. A cricothyrotomy may be necessary to maintain the airway, but landmarks may be even more difficult to identify. [Krupa, D. (Ed.). (1997). Flight nursing core curriculum. Park Ridge, IL: National Flight Nurses Association. (pp. 310-311).]

5. **d**
Securing an airway is the first priority in any patient. In a patient with anterior neck trauma the airway is more likely to be a problem requiring intervention. [Krupa, D. (Ed.). (1997). Flight nursing core curriculum. Park Ridge, IL: National Flight Nurses Association. (pp. 286).]

6. **d**
Signs and symptoms of an air embolism include hypotension, tachypnea, decreased level of consciousness, and tachycardia. [Krupa, D. (Ed.). (1997). Flight nursing core curriculum. Park Ridge, IL: National Flight Nurses Association. (pp. 309).]

7. d
Effective airway clearance is evidenced by patent airway, bilateral and equal chest excursions with ventilations, and bilateral equal breath sounds. [Krupa, D. (Ed.). (1997). Flight nursing core curriculum. Park Ridge, IL: National Flight Nurses Association. (pp. 311).]

8. **c**
Subcutaneous emphysema and flattened (not enhanced) cricothyroid prominence are objective signs of a fractured larynx. Dysphagia and complaint of pain can occur with a fractured larynx, however, it is a subjective assessment. [Krupa, D. (Ed.). (1997). Flight nursing core curriculum. Park Ridge, IL: National Flight Nurses Association. (pp. 309-310).]

9. **d**
An esophageal injury may cause blood draining into the stomach and could cause vomiting and a possible ineffective airway clearance. The potential for infection is high with substances from the esophagus draining into the thorax. Pain can occur on swallowing, dyspnea, or irritation of the pleural lining. [Holleran, R. (Ed.). (1996). Flight nursing principles and practice. (2nd ed.). Park Ridge, IL: National Flight Nurses Association. (pp. 330).]

10. **d**

The palpation of the neck of a patient with a suspected fractured larynx may reveal subcutaneous emphysema. A flattened cricothyroid prominence may indicate a fractured larynx. The trachea is generally still midline and neck veins are flat. [Krupa, D. (Ed.). (1997). Flight nursing core curriculum. Park Ridge, IL: National Flight Nurses Association. (pp. 309).]

11. **a**

Pupil constriction. CN II – visual fields, CN IV – inferior and medial eye movement and CN VII – eyelid closing. [Krupa, D. (Ed.). (1997). Flight nursing core curriculum. Park Ridge, IL: National Flight Nurses Association. (pp. 282-283).]

12. **b**

A Le Fort II fracture is also known as the pyramidal fracture as it involves a triangular shaped segment of the mid-portion of the face and nasal bones. [Krupa, D. (Ed.). (1997). Flight nursing core curriculum. Park Ridge, IL: National Flight Nurses Association. (pp. 296).]

13. **a**

A hyphema is a collection of blood in the anterior chamber. A retinal detachment may be the cause of the hyphema. [Krupa, D. (Ed.). (1997). Flight nursing core curriculum. Park Ridge, IL: National Flight Nurses Association. (pp. 325).]

Abdominal, Genitourinary, and Gynecological Transport

1. Abdominal gas may expand during ascent as a result of:

    a. a ruptured diaphragm.
    b. a displaced endotracheal tube.
    c. progressive shock or anemia.
    d. changes in barometric pressure.

2. Ecchymosis over the flank and abdominal region indicates which of the following conditions?

    a. Fractured pelvis
    b. Ruptured spleen
    c. Lacerated duodenum
    d. Peritoneal hemorrhage

3. Ecchymosis around the umbilicus indicates which of the following conditions?

    a. Appendicitis
    b. Ruptured kidney
    c. Intra-abdominal hemorrhage
    d. Fracture of the ninth and tenth ribs

4. Dullness on percussion of the right upper quadrant most commonly indicates:

    a. a normal finding.
    b. advanced chronic cirrhosis.
    c. free air below the diaphragm.
    d. decreased blood flow through the hepatic artery.

5. If a patient's condition does not improve after nursing interventions, the flight crew should:

    a. continue current interventions until the desired results are attained.
    b. cease current interventions and await further orders.
    c. alter the plan of care and reassess the patient.
    d. none of the above.

6. When transporting a patient with esophageal varices, the flight nurse should not:

    a. insert a nasogastric tube.
    b. initiate blood replacement.
    c. elevate the head of the stretcher.
    d. provide pain management with narcotics.

7. Which of the following medications should not be given to a patient with pancreatitis?

    a. Morphine sulfate
    b. Ketorolac tromethamine
    c. Meperidine hydrochloride
    d. Hydromorphine hydrochloride

8. Assessment of a patient with an injury to the diaphragm is least likely to reveal:

   a. decreased breath sounds.
   b. a persistent air leak in the chest tubes.
   c. audible abdominal peristalsis in the thoracic cavity.
   d. subcutaneous emphysema in the thoracic cavity and abdominal region.

9. You are transporting a 14-year-old skier. Your assessment reveals fixed dullness to percussion in the left flank and dullness in the right flank that disappears when position is changed. You know this as:

   a. Murphy's sign.
   b. Hill's sign.
   c. Ballance sign.
   d. Dressler's syndrome.

10. The transport crew receives report on a patient who has Halstead's sign. The crew knows this is:

   a. bruising of the scrotum or labia.
   b. marbled appearance of the abdomen.
   c. pain in the flank area.
   d. left lower quad pain.

11. Patients may identify kidney pain in which of the following regions?

   a. Suprapubic area
   b. Periumbilical area
   c. Rectum or perineum
   d Area of the costovertebral angle

12. Which of the following conditions is considered a prerenal cause of acute renal failure?

   a. Hypovolemia
   b. Toxic vascular thrombosis
   c. Renal vascular thrombosis
   d. Upper urinary tract obstruction

13. Which of the following is the most appropriate nursing diagnosis   for a patient with acute pyelonephritis?

   a. Potential for impaired skin integrity
   b. Pain related to decreased blood flow and swelling
   c. Altered renal tissue perfusion secondary to trauma
   d. Fluid volume deficit secondary to nausea/vomiting and fever

14. You are transporting a 16-year-old male with severe pain in his lower abdomen that radiates to the inguinal canal. The pain started about 45 minutes ago. VS: P-140, RR-32, BP 146/98. He denies trauma and is tearful. You suspect:

   a. epididymitis.
   b. sexual transmitted disease.
   c. kidney stone.
   d. testicular torsion.

15. A 22-year-old male was involved in a MVC. The exam revealed a positive Grey Turner's sign. The transport nurse should expect what type of injuries?

    a. Bladder rupture
    b. Retroperitoneal hematoma from a pelvic fracture
    c. Liver fracture
    d. Spleen fracture

16. The hallmark sign for GU trauma is:

    a. hematuria.
    b. flank pain.
    c. priapism.
    d. oliguria.

17. Acute renal failure is the loss of normal renal function and the accumulation of metabolic waste products. What is the best indicator and hallmark sign for acute renal failure?

    a. Elevated Serum Creatinine level
    b. Elevated Urine Creatinine level
    c. Elevated Blood Urea Nitrogen (BUN)
    d. Urine output less than 600ml/24 hours

18. Nursing diagnosis for a patient with pelvic inflammatory disease (PID) would include all except:

    a. pain related to inflammation.
    b. anxiety/fear related to pain.
    c. knowledge deficit related to PID risk and disease process.
    d. pain related to vaginal irritation.

Abdominal, Genitourinary, and Gynecological Transport

1. d
Changes in barometric pressure may cause gas expansion and produce pain and discomfort. [Krupa, D. (Ed.). (1997). Flight nursing core curriculum. Park Ridge, IL: National Flight Nurses Association. (pp. 397).]

2. d
Grey Turner's sign indicates peritoneal hemorrhage. [Krupa, D. (Ed.). (1997). Flight nursing core curriculum. Park Ridge, IL: National Flight Nurses Association. (pp. 398).]

3. c
Ecchymosis around the umbilicus indicates intra-abdominal hemorrhage. [Krupa, D. (Ed.). (1997). Flight nursing core curriculum. Park Ridge, IL: National Flight Nurses Association. (pp. 398).]

4. b
 A decrease or absence of dullness over the liver occurs when there is free air under the diaphragm from perforation of the hollow viscous. The liver is displaced downwards causing dullness with abdominal percussion. [Krupa, D. (Ed.). (1997). Flight nursing core curriculum. Park Ridge, IL: National Flight Nurses Association. (pp. 399).]

5. c
In the evaluation component of the nursing process, if the intervention is not effective, the plan is altered to attempt to achieve a desired result. [Krupa, D. (Ed.). (1997). Flight nursing core curriculum. Park Ridge, IL: National Flight Nurses Association. (pp.816).]

6. a
Nasogastric tubes should not be placed in patients with esophageal varices. [Krupa, D. (Ed.). (1997). Flight nursing core curriculum. Park Ridge, IL: National Flight Nurses Association. (pp. 402).]

7. a
Morphine may increase pain due to causing spasms of the Sphincter of Odi. Hydromorphine hydrochloride and Meperidine are the drugs of choice for treatment of acute pancreatitis. [Krupa, D. (Ed.). (1997). Flight nursing core curriculum. Park Ridge, IL: National Flight Nurses Association. (pp. 402).]

8. d
Decreased breath sounds, persistent air leak in chest tubes, and audible abdominal peristalsis in the thoracic cavity are symptoms seen with diaphragmatic injury. [Krupa, D. (Ed.). (1997). Flight nursing core curriculum. Park Ridge, IL: National Flight Nurses Association. (pp. 403).]

9. c
Ballance sign indicates a spleen rupture. Dressler's syndrome seen in post MI. Murphy's sign is the inability to take deep breaths during palpation beneath the RCM. Hill's sign is a disproportionate femoral systolic hypertension. [Krupa, D. (Ed.). (1997). Flight nursing core curriculum. Park Ridge, IL: National Flight Nurses Association. (pp. 406).]

10. b
A marbled appearance of the abdomen is called Halstead sign. It is seen with Abdominal trauma. [Alspach, J. G. (Ed). (1998) Core curriculum for critical care nursing. (5th ed.). Philadelphia: Saunders. (pp. 708).]

11. d

Renal pain is generally found in the area of the costovertebral angle or the flank area. Suprapubic pain is associated with urethral pain. Bladder inflammation is felt as a sharp burning pain at the urethral tip. Prostate pain is described as a discomfort in the lower back, rectum or perineum. [Krupa, D. (Ed.). (1997). Flight nursing core curriculum. Park Ridge, IL: National Flight Nurses Association. (pp. 446-447).]

12. a

Prerenal causes of acute renal failure (ARF) are due to a decrease in blood flow to the kidneys hypovolemia is an example of this. Toxic vascular thrombosis and renal vascular thrombosis are problems occurring within the kidney and are considered renal causes of ARF. Upper urinary tract obstruction occurs outside the kidney in the urinary tract system and is distal to the kidney making this a postrenal cause of ARF. [Krupa, D. (Ed.). (1997). Flight nursing core curriculum. Park Ridge, IL: National Flight Nurses Association. (pp. 450).]

13. d

Acute pyelonephritis is an infectious inflammatory disease in which fever, nausea, and vomiting occur; these signs and symptoms can result in possible fluid volume deficit. Impaired skin integrity is a potential collaborative problem with acute renal failure not pyelonephritis. Decreased blood flow is not a usual problem with pyelonephritis and the disease is due to infection not trauma. [Krupa, D. (Ed.). (1997). Flight nursing core curriculum. Park Ridge, IL: National Flight Nurses Association. (pp. 452 - 453).]

14. d

Severe onset of pain, which may radiate to lower abdomen and inguinal canal are signs of torsion. Additional signs include scrotal swelling, and an elevated testes on the affected side. [Krupa, D. (Ed.). (1997). Flight nursing core curriculum. Park Ridge, IL: National Flight Nurses Association. (pp. 454).]

15. b

Grey Turner's sign is bruising at the flank area and is indicative of a retroperitoneal hematoma usually associated with pelvic fractures. [Krupa, D. (Ed.). (1997). Flight nursing core curriculum. Park Ridge, IL: National Flight Nurses Association. (pp. 398).]

16. a

Hematuria is the hallmark sign of GU trauma. Patients with GU trauma have small to large amounts of blood in the urine 80% of the time. Microscopic blood is present in the remaining 20%. All trauma patients should have a urinalysis to detect microscopic RBCs. [Krupa, D. (Ed.). (1997). Flight nursing core curriculum. Park Ridge, IL: National Flight Nurses Association. (pp. 456-457).]

17. d

The best indicator for ARF is decrease in urinary output. Elevation of serum creatinine occurs as by-products rise in the serum due to the decrease in GFR but a decrease in urine output is the hallmark sign of ARF. Elevation of BUN can be from gastrointestinal bleeding and cause a falsely high BUN. [Alspach, J. G. (Ed). (1998) Core curriculum for critical care nursing. (5th ed.). Philadelphia: Saunders. (pp. 499-507).]

18. d

Pain is caused by the inflammation & infection. [Krupa, D. (Ed.). (1997). Flight nursing core curriculum. Park Ridge, IL: National Flight Nurses Association. (pp. 463).]

19. b

Ectopic pregnancy presents with lower abdomen pain, usually unilateral, cramping, aggravated by motion, may have referred shoulder pain, N/V, dizziness, and possible spotting. [Krupa, D. (Ed.). (1997). Flight nursing core curriculum. Park Ridge, IL: National Flight Nurses Association. (pp. 461).]

General Medical Transport

1. Fluid loss in a dehydrated patient will most critically increase serum levels of which of the following?

    a. Sodium
    b. Calcium
    c. Chloride
    d. Potassium

2. Medical patients with electrolyte imbalances who are being transported should be regularly assessed and monitored for changes in:

    a. level of consciousness.
    b. cardiac rhythm.
    c. hydration status.
    d. all of the above.

3. Expected outcomes for a well-managed patient being transported for electrolyte   imbalances include:

    a. focal seizure activity.
    b. normal pulse and blood pressure.
    c. Glasgow Coma Scale score of 12.
    d. urine output of less than 30 mL/hour.

4. Which of the following interventions is not indicated when transporting a patient with syndrome of inappropriate antidiuretic hormone (SIADH)?

    a. Fluid restriction
    b. Diuretics as ordered
    c. Insertion of an indwelling urinary catheter
    d. Administering an IV bolus of 5% dextrose in water

5. Which of the following medical conditions is most likely to result in hypoglycemia?

    a. Alcoholism
    b. Hyponatremia
    c. Von Willebrand's disease
    d. Disseminated intravascular coagulation (DIC)

6. Intervention for a patient with signs of hypoglycemia should include:

    a. administration of insulin.
    b. IV administration of D10W.
    c. IV administration of potassium.
    d. administration of fluid boluses with 0.9% normal saline.

7. Potential in-flight interventions for a patient with diabetic ketoacidosis (DKA) include:

    a. administration of D10W.
    b. fluid restriction to less than 2,000 mL/day.
    c. administration of insulin and IV saline boluses.
    d. oral administration of high glucose supplements.

8. What is the best way for a flight nurse to monitor hydration status in a patient experiencing thyrotoxic crisis during transport?

    a. Frequent assessment of skin turgor
    b. Close monitoring of IV, oral intake, and urinary and nasogastric output
    c. Use of the Parkland formula, with compensation for the decreased humidity in the aircraft
    d. Monitoring of hydration status is not necessary, as hydration status is not altered in patients with this condition

9. The expected outcome of interventions for a patient with dehydration is:

    a. no seizure activity.
    b. dry mucous membranes.
    c. capillary refill of 3 to 4 seconds.
    d. blood pressure of 80/60 mm Hg, heart rate of 128 beats/mm, and respirations of 24/mm.

10. When transporting medical patients from a rural facility, the air medical flight team should:

    a. repeat all essential laboratory studies on arrival to obtain pre-transport baseline.
    b. wait for the unit secretary to copy the patient's entire medical record before leaving the referring hospital.
    c. encourage that the referring hospital faxes any pending laboratory results to the receiving facility as soon as possible.
    d. disregard the rural physician's findings, as these physicians are frequently inexperienced at managing complex medical problems.

11. Which of the following arterial blood gas values would you suspect to see in a patient that is diagnosed with DKA?

	PH	PaO2	PaCO2	HCO3
a.	7.40	80	30	22
b.	7.40	70	22	33
c.	7.27	90	50	20
d.	7.20	88	23	16

12. The patient is uncooperative, HR 136, RR-24, BP 90/53. The patient has a history of renal failure, dialysis, IV drug abuse, and diet controlled diabetes. You suspect part of the problem causing the patient's agitation to be related to:

    a. hypoglycemia.
    b. diabetic ketoacidosis (DKA).
    c. hyperglycemic hyperosmolar nonketotic coma (HHNC).
    d. adrenal crisis.

General Medical Transport Answers

1. a

Dehydration is a loss of water without the loss of sodium, thereby causing an increase in sodium. The cause of the fluid loss may create other electrolyte abnormalities, but not to the extent of sodium. [Krupa, D. (Ed.). (1997). Flight nursing core curriculum. Park Ridge, IL: National Flight Nurses Association. (pp. 342).]

2. d

Medical patients being transported with electrolyte abnormalities will need to be assessed and monitored for changes in level of consciousness from lethargy to coma, changes in cardiac rhythm due primarily to changes in sodium and potassium, and alterations in hydration status, (i.e., skin turgor and dry mucous membranes). [Krupa, D. (Ed.). (1997). Flight nursing core curriculum. Park Ridge, IL: National Flight Nurses Association. (pp. 343).]

3. b

Normal blood pressure and pulse rates are expected outcomes of appropriate treatment of patients with electrolyte imbalance. Focal seizures, decreased level of consciousness, and decreased urine output may be occurring, but are indications of continued problems requiring intervention. [Krupa, D. (Ed.). (1997). Flight nursing core curriculum. Park Ridge, IL: National Flight Nurses Association. (pp. 344).]

4. d

Intravenous fluids administered to patients with syndrome of inappropriate antidiuretic hormone (SIADH) need to be hypertonic solutions, not solutions with free water in them. Interventions for patients with SIADH do include fluid restriction, (especially free fluids) insertion of an indwelling urinary catheter for urine output measurement, and administration of diuretics as ordered and needed. [Krupa, D. (Ed.). (1997). Flight nursing core curriculum. Park Ridge, IL: National Flight Nurses Association. (pp. 350).]

5. a

The primary causes of hypoglycemia are: diabetes with dietary intake and insulin dose mismatches, and alcohol consumption. DIC and Von Willebrand's disease are coagulopathies, which do not generally have a resultant hypoglycemia. Hypernatremia is the sodium abnormality associated with hypoglycemia. [Krupa, D. (Ed.). (1997). Flight nursing core curriculum. Park Ridge, IL: National Flight Nurses Association. (pp. 344, 366).]

6. b

For administration of rapidly available glucose, D10W is an intervention that may occur in flight. Insulin, potassium and fluid boluses of normal saline are appropriate interventions in patients with hyperglycemia. [Krupa, D. (Ed.). (1997). Flight nursing core curriculum. Park Ridge, IL: National Flight Nurses Association. (pp. 369, 372).]

7. c

The treatment of DKA includes administration of insulin and intravenous saline boluses. The patient already has an increased glucose level so the administration of dextrose would be inappropriate whether intravenous or oral. [Krupa, D. (Ed.). (1997). Flight nursing core curriculum. Park Ridge, IL: National Flight Nurses Association. (pp. 372-373).]

8. b

Close monitoring of intake and output is critical to determine appropriate interventions. These patients can have hypertension and congestive heart failure or present with hypotension and cardiovascular collapse. The Parkland formula for fluid resuscitation is related to burn patients most commonly, not thyrotoxic crisis. Skin turgor is not the most accurate method of assessing for hydration status, which is important considering the problems these patients can encounter. [Alspach, J. (Ed.). (1996). Core curriculum for critical care nursing. (5th ed.). Philadelphia: Saunders. (pp. 637 - 644).]

9. a

The expected outcomes of interventions for a dehydrated patient include no seizure activity, vital signs within normal limits, and no signs of dehydration. [Krupa, D. (Ed.). (1997). Flight nursing core curriculum. Park Ridge, IL: National Flight Nurses Association. (pp. 344).]

10. c

Transport should not be delayed awaiting lab work results. [Krupa, D. (Ed.). (1997). Flight nursing core curriculum. Park Ridge, IL: National Flight Nurses Association. (pp. 340).]

11. d

A patient with diabetic ketoacidosis (DKA) you would suspect to see a metabolic acidosis and a respiratory alkalosis. [Krupa, D. (Ed.). (1997). Flight nursing core curriculum. Park Ridge, IL: National Flight Nurses Association. (pp. 372).]

12. c

HHNC usually affect type II diabetics. Assessment of HHNC includes hypotension, glucose > 800, tachycardia, normal respirations, altered LOC, alertness, dehydration, dry skin, possible fever & seizures. [Krupa, D. (Ed.). (1997). Flight nursing core curriculum. Park Ridge, IL: National Flight Nurses Association. (pp. 373-374).]

Neurological Transport

1. Appropriate outcome(s) for a patient who has had a cerebrovascular accident include:

    a. ability to communicate.
    b. effective airway clearance.
    c. ability to sustain spontaneous ventilation.
    d. all of the above.

2. A woman is unable to flex her hips following a head-on motor vehicle crash in which she was secured only by a lap belt. This finding suggests an injury at the level of:

    a. T7 - T8.
    b. T9 – T10.
    c. L1 - L2.
    d. L3 - L4.

3. When transporting the patient with a spinal cord injury, hanging weights for cervical traction should be:

    a. increased by 1/3 due to altitude considerations.
    b. decreased by 1/3 due to altitude considerations.
    c. maintained during transport to prevent further cord injury.
    d. avoided, as they are inappropriate in the air medical setting.

4. Intervention by the transport team for a patient with a spinal cord injury and autonomic dysreflexia should include:

    a. consideration of altitude adjustment.
    b. administering an antihypertensive agent.
    c. administering an anxiolytic or analgesic agent.
    d. confirming the patency of urinary drainage catheter.

5. A patient with meningitis is being transported from a community hospital to a larger facility for definitive care. The air medical team should use respiratory personal protective equipment if the patient has what type of meningitis?

    a. Viral
    b. Fungal
    c. Aseptic
    d. Bacterial

6. Of the following, which treatment sequence is most appropriate for a patient with increased intracranial pressure during transport?

    a. Elevate the head of the bed, sedate the patient, and give mannitol
    b. Elevate the head of the bed, give mannitol, and sedate the patient
    c. Give mannitol, elevate the head of the bed, and sedate the patient
    d. Provide IV sedation, give mannitol, and elevate the head of the bed

7.	Treatment of increased intracranial pressure with artificial ventilation can be done during transport by:

	a.	decreasing the tidal volume.
	b.	increasing the respiratory rate.
	c.	decreasing the respiratory rate.
	d.	increasing the ratio of inspirations to expirations.

8.	When transporting the patient with an acute hemorrhagic stroke, blood pressure should be maintained:

	a.	slightly above the patient's norm.
	b.	slightly below the patients norm.
	c.	significantly below the patient's norm.
	d.	none of the above.

9.	A patient who has a 6-inch spike embedded in his temporal bone has a Glasgow Coma Scale score of 15 with no neurologic deficits. The patient is immobilized on a backboard in full spinal precautions. The most appropriate course of action would be:

	a.	intubate and prepare the patient for transport.
	b.	secure the object and prepare the patient for transport.
	c.	refuse air transport based on the instability of the injury.
	d.	remove the object and place a pressure dressing over the skull prior to transport.

10.	The best method of monitoring a head injured patient during transport is to regularly assess the patients:

	a.	spinal reflexes.
	b.	pupillary response.
	c.	respiratory pattern.
	d.	level of consciousness.

11.	A patient being transported following a motor vehicle crash suddenly becomes obtunded and withdraws to painful stimuli only. The transport team should first:

	a.	prepare for intubation.
	b.	administer 25 g of IV mannitol.
	c.	administer an analgesic for pain.
	d.	elevate the head of the bed to reduce intracranial pressure.

12.	A patient with a subarachnoid bleed due to rupture of an aneurysm is being transported when he reports a severe headache. The most appropriate treatment would be to:

	a.	administer an anxiolytic agent.
	b.	administer analgesics as ordered.
	c.	administer an antihypertensive agent.
	d.	provide verbal reassurance and minimize external stimuli.

13. Cushing's response is demonstrated by a:

    a. rising systolic blood pressure, widening pulse pressure, and bradycardia.
    b. rising systolic blood pressure, narrowing pulse pressure, and tachycardia.
    c. decreasing systolic blood pressure, narrowing pulse pressure, and tachycardia.
    d. decreasing systolic blood pressure, widening pulse pressure, and bradycardia.

14. A positive Brudzinski's sign in a conscious patient (immediate flexion of the hips and knees when the head is raised off the pillow) is symptomatic of:

    a. a seizure disorder.
    b. meningeal irritation.
    c. a thromboembolic event.
    d. an aneurysm with a bleed.

15. A patient with a history of transient loss of consciousness, followed by a period of lucidity, then a second period of decreasing LOC most likely has:

    a. epidural bleed.
    b. subarachnoid bleed.
    c. acute subdural bleed.
    d. chronic subdural bleed.

16. A potential nursing diagnosis/collaborative problem for a patient with a cerebral aneurysm is:

    a. fluid volume excess.
    b. high risk for infection.
    c. impaired skin integrity.
    d. altered cerebral tissue perfusion.

17. Stroke patients are often kept slightly hypertensive in order to:

    a. decrease intracranial pressure.
    b. decrease cerebral perfusion pressure.
    c. maintain cerebral perfusion pressure.
    d. maintain cerebrospinal fluid pressure.

18. During transport of a patient with a cerebral aneurysm who has stable vital signs, appropriate interventions include all the following except:

    a. minimizing pain with analgesics.
    b. maintaining normothermia and preventing shivering.
    c. keeping the head of the bed flat throughout transport.
    d. preventing the patient from performing the Valsalva maneuver.

19. Increased intracranial pressure can develop in a patient with a ventricular shunt if which of the following problems occur?

    a. Kinking
    b. Plugging
    c. Bacterial infection
    d. All of the above problems

20. Assessment of a head injured patient reveals periorbital ecchymosis. This finding provides the transport team with a high index of suspicion for which type of fracture?

   a. Maxillary fracture
   b. Basal skull fracture
   c. Frontal sinus fracture
   d. Unilateral orbital blowout fracture

21. Secondary injury to the spinal cord is typically caused by:

   a. ischemia.
   b. laceration.
   c. contusion.
   d. stretching.

22. A patient with a complete transection of the spinal cord at the level of the sixth cervical vertebra should have an intact Foley catheter during transport to prevent:

   a. bradycardia.
   b. massive vasodilatation.
   c. decreased parasympathetic stimulation.
   d. uncompensated hyperdynamic cardiovascular response.

23. Effective airway clearance in a neurologically impaired patient is evident during transport by:

   a. absence of emesis.
   b. equal breath sounds.
   c. the presence of a clear and patent airway.
   d. a pulse oximetry reading of greater than 92%.

24. The transport team knows that the most reliable assessment of a patient's neurological condition is:

   a. LOC.
   b. pupillary response.
   c. vital signs.
   d. motor function.

25. How would the transport nurse know that there is damage to the sixth cranial nerve in a patient with a closed head injury?

   a. Absent gag reflex
   b. Fixed and dilated pupils
   c. Nystagmus
   d. Inward deviation of the eyes

26.   You are preparing to transport a 26-year-old construction worker who fell 20 feet from a scaffold. He has greater motor loss in his upper extremities than in his lower, with varying sensory loss. The team recognizes this as:

a.   complete transection.
b.   central cord syndrome.
c.   Brown-Sequard syndrome.
d.   anterior cord syndrome.

27.   You are transporting a 32 year old who fell from a horse. During your assessment it is noted that the patient cannot shrug his shoulders. Which cranial nerve (CN) affects this?

a.   CN XII
b.   CN XI
c.   CN VII
d.   CN V

Neurological Transport Answers

1. d
CVA patients are at risk for problems with airway, ventilation and communication. Outcomes of appropriate treatment would include the patient's ability to effectively keep the airway clear, sustain spontaneous ventilation, and communicate. [Krupa, D. (Ed.). (1997). Flight nursing core curriculum. Park Ridge, IL: National Flight Nurses Association. (pp. 425).]

2. c
An L 1 -2 injury results in neural impairment of the iliopsoas muscles of the hip, which are responsible for hyperflexion. [Holleran, R. (Ed.). (2003). Air and surface patient transport: Principles and practice. (3rd ed.). St. Louis: Mosby. (pp. 273).]

3. d
Hanging weights are considered inappropriate in the air medical setting because they are not secured during flight and are a safety hazard. They may be secured to the stretcher after consultation with a neurosurgeon. Altitude does not affect traction weight. [Krupa, D. (Ed.). (1997). Flight nursing core curriculum. Park Ridge, IL: National Flight Nurses Association. (pp. 440).]

4. d
Although administration of an antihypertensive agent may be indicated, the nurse should first confirm patency of the urinary drainage catheter since symptoms often abate once the stimulus is removed. Altitude adjustment does not affect symptoms of autonomic dysreflexia, and anxiolytics or analgesics are not indicated in this syndrome of sympathetic hyper responsiveness. [Krupa, D. (Ed.). (1997). Flight nursing core curriculum. Park Ridge, IL: National Flight Nurses Association. (pp. 440).]

5. d
Bacterial meningitis, requires use of a mask during transport in a confined space due to possibility of droplet source of infection. [Holleran, R. (Ed.). (2003). Air and surface patient transport Principles and practice. (3rd ed.). St. Louis: Mosby. (pp. 451).]

6. a
The most appropriate sequence includes starting with the intervention that has the least potentially harmful effects to the patient. All three of the interventions are correct, and elevating the head of the bed would be the least invasive as a first attempt to reduce increased intracranial pressure. Sedation would be a more invasive method if the patient is agitated, while mannitol would be the intervention that would be implemented if the first two did not decrease the intracranial pressure. [Krupa, D. (Ed.). (1997). Flight nursing core curriculum. Park Ridge, IL: National Flight Nurses Association. (pp. 442).]

7. b
Intracranial pressure can be reduced by decreasing arterial CO_2, which acts as a vasodilator in the cerebral blood vessels. Decreasing arterial CO_2 can be accomplished by increasing the respiratory rate or increasing the tidal volume on a mechanically ventilated patient. Increasing the inspiratory to expiratory ratio would improve oxygenation but would not decrease arterial CO_2. [Krupa, D. (Ed.). (1997). Flight nursing core curriculum. Park Ridge, IL: National Flight Nurses Association. (pp. 421).]

8. a
Blood pressure should be maintained slightly above the patient's norm to maintain cerebral perfusion pressure. [Krupa, D. (Ed.). (1997). Flight nursing core curriculum. Park Ridge, IL: National Flight Nurses Association. (pp. 426).]

9. b
Securing the object and preparing for transport is the most appropriate action. Removing the object would create a much worse bleed that may be successfully tamponade by the object while it is embedded. Since the patient is alert and without deficits, intubation is not necessary for transport. Refusal of air transport based on the injury is inappropriate since the object can be secured and there are no safety concerns. [Krupa, D. (Ed.). (1997). Flight nursing core curriculum. Park Ridge, IL: National Flight Nurses Association. (pp. 439).]

10. d
A change in the level of consciousness is the first sign of worsening of brain injury and can be assessed throughout transport by assessing patient responsiveness to verbal or painful stimuli. Spinal reflexes have limited usefulness in an air medical setting or with a head injured patient. A change in pupillary response or respiratory pattern is late changes in head injured patients. [Holleran, R. (Ed.). (2003). Air and surface patient transport: Principles and practice. (3rd ed.). St. Louis: Mosby. (pp. 263).]

11. a
Patient care should always begin with the ABC assessment. Although it is likely that this patient has increased intracranial pressure and may benefit from elevating the head of the trauma patient and mannitol, airway management is always the priority. Administering an analgesic for pain is not indicated and may further depress respirations before an airway is obtained. [Krupa, D. (Ed.). (1997). Flight nursing core curriculum. Park Ridge, IL: National Flight Nurses Association. (pp. 417).]

12. b
Patients with subarachnoid bleeds historically have severe headaches. Not only is an analgesic appropriate treatment for this patient but will also be helpful in reducing anxiety and maintaining a controlled blood pressure. Anxiolytics and antihypertensive agents have no analgesic effect and will be of limited benefit in controlling headache pain. Although verbal reassurance and minimized external stimuli may help, an analgesic is the most definitive therapy for pain control. [Krupa, D. (Ed.). (1997). Flight nursing core curriculum. Park Ridge, IL: National Flight Nurses Association. (pp. 427).]

13. a
Cushing's response is a very late finding in a neurologic patient with increased intracranial pressure and results in a rising systolic blood pressure, widening pulse pressure and bradycardia. [Krupa, D. (Ed.). (1997). Flight nursing core curriculum. Park Ridge, IL: National Flight Nurses Association. (pp. 414)].

14. b
Signs of meningeal irritation include Brudzinski's sign, Kernig's sign, projectile vomiting, stiff neck, photophobia, and blurred vision. Brudzinski's is the immediate flexion of the knees and hips when the head is raised off the pillow. [Krupa, D. (Ed.). (1997). Flight nursing core curriculum. Park Ridge, IL: National Flight Nurses Association. (pp. 429)].

15. a
An epidural bleed has a unique LOC pattern, with a transient loss of consciousness, followed by lucidity, then a decreasing level of consciousness. Subarachnoid bleeds generally have a sudden loss of consciousness following a sudden increase in intracranial pressure. Acute subdural bleeds and chronic subdural bleeds generally have a slower onset of decreased level of consciousness. [Krupa, D. (Ed.). (1997). Flight nursing core curriculum. Park Ridge, IL: National Flight Nurses Association. (pp. 435)].

16. d
Cerebral aneurysms have the potential for rupture and thereby causing an area of decreased cerebral tissue perfusion. A cerebral aneurysm does not cause an overall fluid volume excess. Infection and skin integrity are not usual problems in an emergent setting with cerebral aneurysms. [Krupa, D. (Ed.). (1997). Flight nursing core curriculum. Park Ridge, IL: National Flight Nurses Association. (pp. 426 -427)].

17. c
Stroke patients are kept hypertensive to maintain cerebral perfusion pressure. Hypertension will increase intracranial pressure, not decrease it. [Krupa, D. (Ed.). (1997). Flight nursing core curriculum. Park Ridge, IL: National Flight Nurses Association. (pp. 426)].

18. c
If the patient is not a trauma patient keeping the head of the bed elevated is an appropriate intervention to assist in decreasing blood pressure on the aneurysm site. Aneurysms may rupture with a sudden increase in cerebral pressure, like during a Valsalva maneuver. Anything that increases basal metabolic rate is to be avoided; e.g., shivering and pain. [Krupa, D. (Ed.). (1997). Flight nursing core curriculum. Park Ridge, IL: National Flight Nurses Association. (pp. 427)].

19. d
A ventricular shunt can increase intracranial pressure if it becomes kinked, plugged, or a bacterial infection causes the shunt to stop functioning and draining. [Krupa, D. (Ed.). (1997). Flight nursing core curriculum. Park Ridge, IL: National Flight Nurses Association. (pp. 431)].

20. b
Raccoon's eyes (periorbital ecchymosis) occurs within several hours of a basal skull fracture. [Krupa, D. (Ed.). (1997). Flight nursing core curriculum. Park Ridge, IL: National Flight Nurses Association. (pp. 433)].

21. a
Ischemia is a secondary injury that can occur in a spinal cord injury. Laceration, contusion and stretching are some of the primary causes of damage to the spinal cord. [Krupa, D. (Ed.). (1997). Flight nursing core curriculum. Park Ridge, IL: National Flight Nurses Association. (pp. 439)].

22. d
A foley catheter to keep the bladder empty is important to prevent a stimulation of the sympathetic nervous system below the level of the spinal cord lesion. This can cause a massive uncompensated hyperdynamic cardiovascular response that can be life threatening. Bradycardia and vasodilatation occur due to the loss or disruption of the autonomic nervous system allowing for unopposed parasympathetic stimulation. [Krupa, D. (Ed.). (1997). Flight nursing core curriculum. Park Ridge, IL: National Flight Nurses Association. (pp. 440)].

23. c
A clear and patent airway is an indication of effective airway clearance. Equal breath sounds indicate the ability to sustain ventilation. Appropriate gas exchange is indicated by oxygen saturation via pulse oximetry of > 92%. The absence of emesis is an indication of effective care for the high risk of aspiration. [Krupa, D. (Ed.). (1997). Flight nursing core curriculum. Park Ridge, IL: National Flight Nurses Association. (pp. 419)].

24. a
A change in the patients LOC is the earliest and most sensitive indicator of a change in neurological status. An altered LOC is the hallmark of cerebral injury. [Krupa, D. (Ed.). (1997). Flight nursing core curriculum. Park Ridge, IL: National Flight Nurses Association. (pp. 419)]

25. d
The Abducens nerve controls movement of the eyes. The gag reflex is controlled by cranial nerve IX and X. Cranial nerve III controls pupillary response. Nystagmus is caused by an interruption of cerebellar function. [Krupa, D. (Ed.). (1997). Flight nursing core curriculum. Park Ridge, IL: National Flight Nurses Association. (pp. 415-416)].

26. b
In central cord syndrome the patient will have greater motor loss in upper extremities than in lower; with varying sensory loss. Complete transection causes loss of sensory and motor function below that lesion that is irreversible. Brown-Sequard syndrome is a hemisection of the cord. The patient will exhibit ipslateral loss of motor, position, and vibratory sense with contralateral loss of pain and temperature sensation. An anterior cord syndrome presents with complete motor loss; loss of pain and temperature below the level of the lesion with sparing proprioception, vibratory sense and touch. [ASTNA. (2002). Transport nurse advance trauma course. (3rd ed.). Denver, Colorado: Air and Surface Transport Nurses Association. (Chapter 9, pg 7)].

27. b
Cranial nerve XI is responsible for head turning and shoulder shrugging. Cranial nerve XII is the tongue. CN VII is responsible for facial movement and expression. CN V corresponds with eye reflexes. [Krupa, D. (Ed.). (1997). Flight nursing core curriculum. Park Ridge, IL: National Flight Nurses Association. (pg. 415-416)].

Obstetric Transport

1. Pre-term rupture of amniotic membranes can be confirmed by all of the following, except:

    a.   positive cervical dilation.
    b.   positive nitrazine testing of fluid.
    c.   pooling of fluid in the vaginal vault.
    d.   positive ferning on microscopic slide.

2. Initial management of a patient with preterm rupture of membranes include giving:

    a.   sublingual nifedipine.
    b.   subcutaneous Terbutaline.
    c.   adequate hydration and providing proper positioning.
    d.   a bolus of magnesium sulfate with a follow-up infusion.

3. What is considered the most recognized primary cause of preterm labor?

    a.   Infection
    b.   Uterine anomalies
    c.   Cigarette smoking
    d.   Lack of prenatal care

4. Beta-sympathomimetic agents, such as Terbutaline or ritodrine are contraindicated in patients who have:

    a.   hypotension.
    b.   gone into preterm labor.
    c.   insulin-dependent diabetes.
    d.   been given a magnesium sulfate infusion.

5. What are the three classic signs and symptoms of pregnancy- induced hypertension?

    a.   Headache, hyperreflexia, and HELLP
    b.   Hypertension, edema, and proteinuria
    c.   Hypertension, headache, and hyperreflexia
    d.   Epigastric pain, hyperreflexia, and proteinuria

6. Clinical signs of placenta previa include all of the following except:

    a.   repetitive and episodic blood loss.
    b.   intense initial abdominal pain.
    c.   vaginal bleeding at the second trimester.
    d.   contractions occurring after the hemorrhage.

7. An expectant mother experiencing vaginal bleeding should not be transferred if:

    a.   she has hypotension or is in shock.
    b.   placenta previa has been ruled out.
    c.   fetal heart tones are 110 beats/mm but reactive.
    d.   she has saturated five pen-pads in the last 10 hours.

8. Which of the following findings is considered a late sign of hypovolemic shock in a pregnant trauma patient?

    a. Respirations of 18 to 20/mm
    b. Urine output of 30 to 50 ml/hour
    c. A maternal heart rate of 100 beats/mm
    d. Fetal heart rate tachycardia, bradycardia, or late decelerations

9. Which of the following maneuvers may be used to help deliver an infant with shoulder dystocia?

    a. Erb's maneuver
    b. Leopold's maneuver
    c. McRobert's maneuver
    d. Forceps technique maneuver

10. Presentation of the lower body parts in the birth canal can cause which of the following complications to the newborn?

    a. Birth asphyxia
    b. Head entrapment
    c. Cord compression
    d. All of the above

11. A 32-year-old woman who has insulin dependent diabetes and is 30 weeks gestation is to be transferred to an obstetrical high-risk center. The patient has had contractions every five minutes for the last hour and has received a one-liter fluid bolus. The flight crew should be prepared to:

    a. administer 0.25 mg of Terbutaline subcutaneously.
    b. maintain an IV of D5W during transport.
    c. check the patient's urine output since the fluid bolus was given.
    d. administer magnesium sulfate 4 mg bolus over 20 to 30 minutes.

12. The expected outcome of interventions for a patient with preterm labor is:

    a. fetal heart rate 80 beats/mm.
    b. slowing or stopping labor.
    c. blood pressure of 160/90 mm Hg, a heart rate of 140 beats/mm, and respirations of 20/mm.
    d. none of the above are expected.

13. Reassuring signs of fetal well-being include:

    a. fetal tachycardia with a less than 5 beat variability.
    b. decrease in variability as labor progresses.
    c. fetal bradycardia of 80-90 beats/mm with a 15 beat variability.
    d. a decrease in fetal heart rate with contractions.

14. Which of the following outcomes are expected following interventions for a pregnant patient with a hypertensive disorder?

    a. Blood pressure of less than 140/90 mm Hg
    b. Fetal heart tones 140 beats/mm
    c. Absences of seizure activity
    d. All of the above

15. Normal characteristics of accelerations of fetal heart rate (FHR) are all of the following except:

    a. they are usually associated with fetal movement.
    b. they can be associated with uterine contractions.
    c. transitory increases above baseline and resemble uterine contractions.
    d. can be indicative of an immature autonomic nervous system.

16. In evaluating fetal heart characteristics what is the most important in determining neurological maturity?

    a. Variability
    b. Flat or decreased beat to beat variability
    c. Accelerations
    d. Transient accelerations and decelerations from baseline FHR

17. The transport team recognizes that DIC is a common complication of:

    a. abruptio placenta.
    b. ectopic pregnancy.
    c. placenta previa.
    d. ovarian rupture.

18. You are transporting a 36-week G-1, P-0 with pre-term labor. Mag Sulfate is infusing at 4 grams/hr. The patient begins to complain of nausea & feeling hot. You should:

    a. reposition patient.
    b. check deep tendon reflexes (DTR).
    c. do a quick cervical exam.
    d. give calcium chloride 1 GM.

19. The first sign of cervical dilation may be:

    a. purulent vaginal discharge.
    b. tachycardia.
    c. vaginal mucus.
    d. increase frequency of contractions.

20. The transport team recognizes that signs of Umbilical Cord Prolapse include:

    a. severe recurrent variable decelerations unresponsive to position.
    b. sudden fetal tachycardia.
    c. maternal complaints of sharp stabbing pain.
    d. boggy enlarge uterus.

Obstetric Transport Answers

1. a
Three findings confirm premature rupture of membranes, including pooling of amniotic fluid in the vaginal vault, positive results on nitrazine testing of fluid, and positive ferning on microscopic slide. [Holleran, R.. (Ed.). (2003). Air and surface patient transport Principles and practice. (3rd ed.). St. Louis: Mosby. (pp. 556).]

2. c
Initial management of a patient experiencing preterm rupture of membranes consists of placing the patient in a left lateral decubitus position to improve uterine perfusion and decrease uterine irritability and providing adequate hydration via a 500 ml bolus of lactated Ringer's or normal saline solution. [Krupa, D. (Ed.). (1997). Flight nursing core curriculum. Park Ridge, IL: National Flight Nurses Association. (pp. 469 -470).] [Holleran, R. (Ed.). (2003). Air and surface patient transport: Principles and practice. (3rd ed.). St. Louis: Mosby. (pp. 557-558).]

3. a
Although many factors predispose pregnant women to preterm labor, few single identifiable causes exist. Infection has been recognized as the primary cause of preterm labor. [Holleran, R. (Ed.). (2003). Air and surface patient transport: Principles and practice. (3rd ed.). St. Louis: Mosby. (pp. 555).]

4. c
Contraindications to the beta-sympathomimetic medication Terbutaline is maternal pulse of greater than 120 beats per minute, suspected chorioamnionitis, insulin-dependent diabetes mellitus, chronic hypertension, or active hemorrhage. [Krupa, D. (Ed.). (1997). Flight nursing core curriculum. Park Ridge, IL: National Flight Nurses Association. (pp. 469-472).]

5. b
The three principal findings to confirm a diagnosis of pregnancy-induced hypertension (PIH) in a pregnant women are hypertension, edema and proteinuria. [Holleran, R. (Ed.). (2003). Air and surface patient transport: Principles and practice. (3rd ed.). St. Louis: Mosby. (pp. 550).]

6. b
Clinical signs of placenta previa include contraction, but they may not initially be present. Abrupt intense abdominal pain may occur, but typically occurs during or after the hemorrhage. Bleeding episodes become repetitive and frequently more extensive. In addition, as the pregnancy progresses, the cervix softens or effaces, which may cause bleeding, usually in the second trimester. [Holleran, R. (Ed.). (2003). Air and surface patient transport: Principles and practice. (3rd ed.). St. Louis: Mosby. (pp. 546).]

7. a
A pregnant woman should not be transported if she has hypotension, is in shock, or is experiencing brisk or active hemorrhaging. In addition, transport is not appropriate if there is evidence of fetal compromise (repetitive late decelerations). [Krupa, D. (Ed.). (1997). Flight nursing core curriculum. Park Ridge, IL: National Flight Nurses Association. (pp. 478).]

8. c
Hypovolemic shock is a common result of trauma. Late signs and symptoms in the mother include a pulse of more than 100 beats/minute, respirations of more than 20/min, and urine output of less than 30ml/hour. Late signs of hypovolemic shock manifested by the fetus include a fetal heart rate that is tachycardic, bradycardic or showing late decelerations. [Krupa, D. (Ed.). (1997). Flight nursing core curriculum. Park Ridge, IL: National Flight Nurses Association. (pp. 480).]

9. c
The McRobert's maneuver, a simple maneuver that increases diameter of the pelvis by stretching the pelvic joints should be tried in situations in which the shoulders are difficult to deliver. With the patient's legs flexed at the knees, the flight nurse should help the patient draw her knees up and toward her chest and continue with gentle downward traction of the head. [Holleran, R.. (Ed.). (1996). Flight nursing: Principles and practice. (2nd ed.). St. Louis: Mosby. (pp. 614).]

10. d
Complications associated with a breech presentation and delivery includes cord prolapse, cord entanglement, cord compression, head entrapment, fetal birth trauma, and birth asphyxia. [Krupa, D. (Ed.). (1997). Flight nursing core curriculum. Park Ridge, IL: National Flight Nurses Association. (pp. 484).]

11. d
The patient should be given an infusion of magnesium sulfate in order to stop the contractions. The mainline intravenous solution used should be lactated Ringer's or normal saline solution, a solution without glucose. Never administer Terbutaline to a patient with insulin-dependent diabetes because of a transient hyperglycemia response seen with that particular drug. [Holleran, R.. (Ed.). (1996). Flight nursing: Principles and practice. (2nd ed.). St. Louis: Mosby. (pp. 619).]

12. b
The goal of interventions for a patient in preterm labor is to slow or stop the labor. [Krupa, D. (Ed.). (1997). Flight nursing core curriculum. Park Ridge, IL: National Flight Nurses Association. (pp. 472).]

13. a
One reassuring sign of fetal well being is a fetal heart beat with less than a five beat variability. Signs that may be a cause of concern about fetal well-being include a significant increase or decrease in the fetal heart rate baseline during a period of several hours, a wandering baseline, a spontaneous decrease in variability or a decrease in variability as labor progresses, bradycardia or tachycardia with reduced variability , subtle late decelerations, or any combination of these signs. [Holleran, R. (Ed.). (2003). Air and surface patient transport: Principles and practice. (3rd ed.). St. Louis: Mosby. (pp. 529).]

14. d
Pregnant patients with hypertensive disorders are at risk for elevated blood pressure, seizure activity, and fetal distress. Successful interventions prevent these complications. [Krupa, D. (Ed.). (1997). Flight nursing core curriculum. Park Ridge, IL.: National Flight Nurses Association. (pp. 474).]

15. d
Spontaneous accelerations of FHR of 15 bpm for 15 seconds in response to feta movement and uterine contractions are indicative of a fetal alertness, well-being, and maturity of the CNS and are a reassuring finding. [Krupa, D. (Ed.). (1997). Flight nursing core curriculum. Park Ridge, IL.: National Flight Nurses Association. (pp. 466-467).]

16. a
Variability of the FHR can be described as the normal irregularity of cardiac rhythm, resulting from a continuous balancing interaction of the sympathetic and parasympathetic branches of the ANS. [Krupa, D. (Ed.). (1997). Flight nursing core curriculum. Park Ridge, IL.: National Flight Nurses Association. (pp. 466-467).]

17. a

A common complication of Abruptio placenta is DIC. Other complications include postpartum hemorrhage, anemia, post partum infection, hypovolemic shock, kidney failure, fetal distress and death. [Holleran, R. (Ed.). (2003). Air and surface patient transport: Principles and practice. (3rd ed.). St. Louis: Mosby. (pp. 545).]

18. b

Signs of Magnesium toxicity include decreased deep tendon reflexes (DTR). Side effects, which can also signal toxicity, include increased warmth, nausea, blurred vision, & respiratory depression. The antidote is calcium gluconate 1 GM of 5% or 10% solution. [Holleran, R. (Ed.). (2003). Air and surface patient transport: Principles and practice. (3rd ed.). St. Louis: Mosby. (pp. 554).]

19. c

Vaginal mucus may be the first sign of cervical dilation. [Holleran, R. (Ed.). (2003). Air and surface patient transport: Principles and practice. (3rd ed.). St. Louis: Mosby. (pp. 557).]20. a Sudden fetal bradycardia and severe recurrent variable decelerations unresponsive to position changes indicate umbilical cord prolapse. [Holleran, R. (Ed.). (2003). Air and surface patient transport: Principles and practice. (3rd ed.). St. Louis: Mosby. (pp. 543).]

20. a

A severe recurrent variable decelerations that are life-threatening for the fetus. [Krupa, D. (Ed.). (1997). Flight nursing core curriculum. Park Ridge, IL.: National Flight Nurses Association. (pp.484).]

Neonatal Transport

1. The high vascular resistance in the fetal lung is due to the following physiologic mechanisms?

    a.   Changes in O_2 tension
    b.   Changes in pH and CO_2 tension
    c.   Pulmonary arterial vasoconstriction
    d.   Increase in systemic vascular resistance

2. Of the following, which group of signs and symptoms is more often associated with neonatal pulmonary disease than with severe congenital heart disease?

    a.   Cyanosis, tachypnea, murmur
    b.   Respiratory distress, hypercarbia, acidosis
    c.   Decreased capillary refill, tachycardia, hypotension
    d.   Decreased peripheral pulses, rales, and hepatomegaly

3. Which of the following findings is the most common predictor of sepsis in a neonate?

    a.   Neutropenia
    b.   Eosinophilia
    c.   Leukocytosis
    d.   Thrombocytopenia

4. You are preparing to transport a newborn. After you have explained to the 17-year-old mother her infants' condition and the need for transport she looks away, does not ask questions, and does not make an attempt to touch her baby. Based on these observations, which of the following nursing diagnoses is not relevant to the mother's reactions?

    a.   Anticipatory grief
    b.   Knowledge deficit
    c.   High risk for injury
    d.   Alteration in family processes

5. Which of the following assessment findings would be a first sign of congestive heart failure in a neonate?

    a.   Pallor
    b.   Heart rate of 160
    c.   Tachypnea
    d.   Rales

6. Which of the following is one of the most common side effects complicating the transport of a neonate receiving prostaglandin therapy?

    a.   Hypertension
    b.   Tachycardia
    c.   Hypoventilation
    d.   Tachypnea

7. In transporting a neonate with severe asphyxia, the transport team recognizes the cause of poor perfusion and mottling in the neonate to be:

   a. septic shock.
   b. hypovolemic shock.
   c. cardiogenic shock.
   d. none of the above.

8. In transporting a neonate, the team suspects the infant has esophageal atresia. What should be an immediate intervention?

   a. Elevate the head of the bed and insert a double lumen suction tube
   b. Intubate
   c. Start an IV and give a fluid bolus
   d. Check glucose

9. A neonate with a defect in the abdominal wall that has other wise completed its development is known as:

   a. omphalocele.
   b. intestinal obstruction.
   c. necrotizing enterocolitis.
   d. gastroschisis.

10. Seizures in the newborn are often confused with:

   a. birth defects.
   b. hypothermia.
   c. jitteriness.
   d. hypoglycemia.

11. During transport of a neonate, which of the following finding would indicate that the neonate is in stress?

   a. Sucking
   b. Fist clinched
   c. Hiccoughing
   d. Tucking

12. During air transport of a neonate with pneumomediastinum the transport team will need to:

   a. increase the SIMV.
   b. increase oxygenation concentrations.
   c. decrease peep.
   d. transport by ground.

13. Which of the following will increase cyanosis in infants with congenital heart disease?

   a. Crying
   b. High O2 concentrations
   c. Fever
   d. Tachypnea

14. In transporting an infant with neural defects which position is optimal?

    a.   Head slightly elevated
    b.   Supine position
    c.   Prone-kneeling position
    d.   Right side

15. Which of the following would not be an intervention used to manage pulmonary   anomalies?

    a.   Monitoring mean airway pressure continuously
    b.   Sedation and paralytics
    c.   Hyperventilation with bag valve mask (BVM)
    d.   Decompression of the bowel

16. In which position should the infant with gastrointestinal anomalies be transported?

    a.   Prone
    b.   Side lying with head slightly elevated
    c.   Supine
    d.   Head up 45°

17. During transport a newborn begins bilious vomiting and his abdomen is distended. Which of the following should the transport team do after securing the airway?

    a.   Gastric decompression
    b.   Fluid bolus with LR
    c.   Check the glucose
    d.   Give medications for comfort

18. Increased irritability, increased HR & BP, eye fluttering, and decrease SaO2 may be subtle signs of:

    a.   hydrocephalus.
    b.   congenital heart defect.
    c.   neurological anomaly.
    d.   seizures.

Neonatal Transport Answers

1. c.
In uteri the primary organ of respiration is the placenta. The lungs are filled with fluid, and the pulmonary arterial vasculature is vasoconstrictor. With the onset of spontaneous respirations at birth, resistance reverses (pulmonary vascular resistance decreases, and systemic vascular resistance increases). [Beachy, P. & Deacon, J. (Eds.). (1993). Core curriculum for neonatal intensive care. Philadelphia: W.B. Saunders. (p. 97).]

2. b.
Signs of respiratory distress, such as retractions, grunting, and flaring, hypercarbia, and a combined respiratory and metabolic acidosis are more indicative of a parenchymal pulmonary disease than of severe congenital heart disease. [Beachy, P. & Deacon, J. (Eds.). (1993). Core curriculum for neonatal intensive care nursing. Philadelphia: W.B. Saunders. (pp. 98-100).]

3. a.
Leukocytosis may be a normal finding in the newborn. Thrombocytopenia may indicate either a viral or bacterial infection, and if severe, may be associated with disseminated intravascular coagulation (a late finding). Eosinophilia is more commonly associated with an allergic response. Neutropenia (<1500/mm3) then becomes the most accurate predictor of infection in the neonate. [Blackburn, S. & Loper, D. (1992). Maternal, fetal, and neonatal physiology: A clinical perspective. Philadelphia: W. B. Saunders. (pp. 468-470).]

4. c.
The mother's reaction suggests an alteration in the family process, as well as a knowledge deficit due to the catastrophic occurrence of events, the severity of the infant's condition, and the overwhelming complexity of the information this mother is being given. Anticipatory grief may be evident as well, in an attempt to rationalize and cope with the "less than perfect baby", or deal with the feelings of perceived guilt. The nursing diagnosis of high risk for injury applies directly to the neonate rather than to the family's response. [Halliday, H., McClure, G., & Reid, M. (1990). Handbook of neonatal intensive care. (3rd ed.). London: Bailliere Tindall. (pp. 242-243).] [Beachy, P. & Deacon, J. (1993). Core curriculum for neonatal intensive care nursing. Philadelphia: W.B. Saunders. (pp. 144 -146).]

5. b Tachypnea and tachycardia are early symptoms of CHF in the neonate. A heart rate of 140 is normal in a neonate. [Holleran, R. (Ed.). (2003). Air and surface patient transport: Principles and practice. (3rd ed.). St. Louis: Mosby. (pp. 576).]

6. c
The most common side effect complicating transport with use of prostaglandin E1 is apnea or hypoventilation. [Holleran, R. (Ed.). (2003). Air and surface patient transport: Principles and practice. (3rd ed.). St. Louis: Mosby. (pp. 577).]

7. c
PPHN is a syndrome characterized by persistent elevated pulmonary vascular resistance resulting in a right-to-left shunt at the ductus arteriosus or the foramen ovale leading to hypoxemia in the presences of a structurally normal heart. [Holleran, R. (Ed.). (2003). Air and surface patient transport: Principles and practice. (3rd ed.). St. Louis: Mosby. (pp. 577).]

8. a
These infants are at high risk for aspiration either from oropharynx refluxing from the upper esophageal pouch or aspiration of gastric contents from the lower tracheoesophageal fistula. Care is immediate elevation of the head and insertion of a double lumen suction tube into the upper esophageal pouch placed on continuous suction. [Holleran, R. (Ed.). (2003). Air and surface patient transport Principles and practice. (3rd ed.). St. Louis: Mosby. 9pp. 577-578).]

9. d
An omphalocele is an arrest of the development of the abdominal wall. Necrotizing enterocolitis is an ischemia of the bowel. [Holleran, R. (Ed.). (2003). Air and surface patient transport Principles and practice. (3rd ed.). St. Louis: Mosby. (pp. 570).]

10. c
Jitteriness is sensitive to stimulus, whereas seizures are not. [Holleran, R. (Ed.). (2003). Air and surface patient transport: Principles and practice. (3rd ed.). St. Louis: Mosby. (pp. 580).)]

11. c
Hiccoughing is a behavioral cue that the neonate is having stress. [Beachy, P. & Deacon, J. (Eds.). (1993). Core curriculum for neonatal intensive care. Philadelphia: W.B. Saunders. (p. 98-100).]

12. b
Pneumomediastinum is rarely treated. If presence is documented the crew will need to increase the O2 concentrations during air transport. [Krupa, D. (Ed.). (1997). Flight nursing core curriculum. Park Ridge, IL: National Flight Nurses Association. (pp. 510).]

13. a
Infants with severe congenital heart disease usually present with one or more symptoms. Cyanosis is one of those symptoms and crying will increase cyanosis in infants with CHD. [Krupa, D. (Ed.). (1997). Flight nursing core curriculum. Park Ridge, IL: National Flight Nurses Association. (pp. 514).]

14. c
Infants with neural defects should be transported in a prone-kneeling position if tolerated. [Krupa, D. (Ed.). (1997). Flight nursing core curriculum. Park Ridge, IL: National Flight Nurses Association. (pp. 519).]

15. c
Treatment of an infant with pulmonary anomalies would include all except use of BVM. It is recommended to avoid BVM ventilations to prevent bowel distention. [Krupa, D. (Ed.). (1997). Flight nursing core curriculum. Park Ridge, IL: National Flight Nurses Association. (pp. 520-522).]

16. b
Proper position is side lying with head of the bed slightly elevated to decrease intestinal vascular compromise and respiratory distress. [Krupa, D. (Ed.). (1997). Flight nursing core curriculum. Park Ridge, IL: National Flight Nurses Association. (pp. 524).]

17. a
Although an IV needs to be established and frequent glucose monitor is important, the team needs to insert an oral gastric tube to prevent secondary aspiration secondary to vomiting especially if transporting by air. The newborn is exhibiting symptoms of malrotation, which is an anomaly of the intestine. [Krupa, D. (Ed.). (1997). Flight nursing core curriculum. Park Ridge, IL: National Flight Nurses Association. (pp. 528).]

18. d
Subtle seizures may present as increased irritability, increased HR and BP, eye fluttering, and decrease in oxygen saturations. [Krupa, D. (Ed.). (1997). Flight nursing core curriculum. Park Ridge, IL: National Flight Nurses Association. (pp. 500).]

Pediatric Transport

1. A 4-year-old girl found lying at the scene of a motor vehicle crash is apneic and has a weak, thready pulse. After securing an airway with cervical spine precautions and providing respiratory support, the next step in the resuscitation process is to insert:

   a. a peripheral intravenous line.
   b. an arterial line.
   c. an intraosseous line.
   d. a central venous line.

2. What type of endotracheal tube would best maintain the airway of a 9-month-old infant?

   a. 3.5, cuffed
   b. 4.5, cuffed
   c. 4.0, uncuffed
   d. 5.0, uncuffed

3. Drugs that can be administered to a pediatric patient through an intraosseous needle include:

   a. epinephrine.
   b. blood products.
   c. sodium bicarbonate.
   d. all of the above.

4. Children younger than 6 months of age have difficulty with thermoregulation due to all of the following factors except:

   a. small body size.
   b. shivering absent or insufficient.
   c. large surface area to body mass ratio.
   d. underdeveloped neuroreceptors for temperature regulation.

5. The air medical team is to transport a 2-year-old child with a known cyanotic heart defect from a rural clinic to a larger facility. The mother reports several changes in the child's normal cardiac status. Which of the following findings is not considered a typical "red flag" for a child with cyanotic heart disease?

   a. Fever
   b. Poor feeding
   c. Increased cyanosis
   d. Decreased responsiveness

6. Intussusception can best be described as:

   a. a blockage of the bowel caused by adhesions.
   b. an area of bowel that telescopes inward upon itself.
   c. an intestinal blockage that may appear to have associated radiologic abdominal calcification.
   d. a condition that is typified by the radiologic appearance of pneumotosis intestinalis or free intra-peritoneal air.

7. Children under the age of 8 years require uncuffed endotracheal tubes because:

    a.   their airway diameter is smaller.
    b.   their cricoid cartilage serves as a functional cuff.
    c.   the cuff causes excessive pressure on the soft cartilaginous tissue.
    d.   the cuff adds increased resistance during insertion of the endotracheal tube.

8. A nursing diagnosis of primary importance in the pediatric patient is:

    a.   altered body temperature, hypothermia.
    b.   fluid volume deficit.
    c.   altered cardiac output, decreased.
    d.   ineffective breathing pattern.

9. A seven-year-old male is brought into the ED by EMS. The child was riding his bike and slipped off hitting his groin area. He is complaining of severe pain, unable to void, and has bright red blood at the meatus. He probably has:

    a.   saddle injury with probable urethral tear.
    b.   pelvic injuries with a Buck's tear.
    c.   scrotal injury with probable testicular torsion.
    d.   biker's saddle crotch.

10. You are transporting an eight month old with a possible cyanotic heart defect. The infant has a bluish hue color. Which of the   following would not be the defect the infant has?

    a.   VSD
    b.   Coarctation of the aorta
    c.   Transposition of the great vessel
    d.   Hypoplastic left ventricle

11. You are preparing to transport a 3 year old with a diagnosis of croup. Upon viewing his x-ray, you would expect to see the:

    a.   Thumbprint sign.
    b.   Steeple sign.
    c.   Ballance sign.
    d.   Croup signature.

12. The transport team administers an Albuterol nebulizer to a child with asthma. The asthma fails to respond. The team decides to give which of the following next:

    a.   Terbutaline Neb .1mg/Kg in 2cc NS.
    b.   Terbutaline .1mg/kg of a 1:1000 solution SQ.
    c.   Atropine Neb .5 to .75 mg/kg in 2cc NS.
    d.   Ipratropium bromide 250 mcg unit dose inhalation.

13. The transport team checks the glucose of a 2 year old with meningitis and 35 mg% is the result. The transport team should administer which of the following:

    a.   IV bolus of dextrose 25 % at 1-2 mg/kg.
    b.   IV bolus of dextrose 25% 3-5mg/ kg.
    c.   IV bolus of dextrose 10% 1-2 mg/kg.
    d.   IV bolus of dextrose 3-5mg/kg.

14. You are transporting a 6-year old child with an injury to the right eye. During transport the child's right pupil is greater than the left. This is due to injury to which cranial nerve (CN)?

    a.   CN I
    b.   CN III
    c.   CN V
    d.   CN X

15. The transport crew arrives on scene to find a 14-year-old who crashed his dirt bike on the motor cross. He is alert & oriented. Airway patent. States he went over the handlebars and landed on his back. He has asymmetric flaccid paralysis and asymmetric loss of reflexes. He can feel pressure. The crew suspects:

    a.   spinal shock.
    b.   complete cord transection.
    c.   incomplete core transection.
    d.   autonomic reflexia.

16. You are preparing to transport a 6-year-old with severe abdominal pain from the local clinic. The child has swelling in his lower abdomen and pain in the inguinal area. After securing the airway, the goal during transport is:

    a.   fluid bolus with NS to hydrate.
    b.   pain medications.
    c.   decompression and evacuation of gastric cavities.
    d.   immobilization of child.

17. During transport of a 12 month old infant with persistent pulmonary hypertension the goal is to:

    a.   reduce pulmonary vascular resistance(PVR).
    b.   increase pulmonary vascular resistance(PVR).
    c.   reduce systemic vascular resistance(SVR).
    d.   increase systemic vascular resistance(SVR).

18. In the above patient the transport team decides to intubate and place the infant on the ventilator. Which of the following is the goal of this therapy?

    a.   PaO2 (> 100), PaCO2 (25-35)
    b.   PaO2 (90-100), PaCO2 (35-45)
    c.   PaO2 (> 100), PaCO2 (35-45)
    d.   PaO2 (90-100), PaCO2 (25-35)

19. While transport a 12 year old with diabetic ketoacidosis the child becomes lethargic and vomits. Which of the following intervention should the team begin:

    a.   give Narcan.
    b.   insert a nasogastric tube.
    c.   give a fluid bolus.
    d.   check the glucose.

20. During transport of a 13-year-old from a MVC scene the transport team notes the patient has a Hamman's sign. This is seen in:

    a.   diaphragmatic rupture.
    b.   tracheobronchial injuries.
    c.   pulmonary contusion.
    d.   tension pneumothorax.

Pediatric Transport Answers

1. a

Peripheral lines are the first intravenous access to be considered. Percutaneous peripheral attempts limited to two attempts. If a peripheral line is not obtained, an intraosseous line should be considered for this child. Intraosseous and central lines are not the first choice for intravenous access in children. [Holleran, R. (Ed.). (2003). Air and surface patient transport: Principles and practice. (3rd ed.). St. Louis: Mosby. (pp. 618).]

2. c

A 4.0, uncuffed endotracheal tube is the most likely size for a nine month old. Uncuffed endotracheal tubes are typically used on children younger than 8-10 years old. [Hazinski, M. (Eds.). (2002). Pediatric advance life support provider manual. American Heart Association. (pp. 107).]

3. d

Medications are well absorbed from the bone marrow; therefore, an intraosseous is considered equivalent to an intravenous. Thus, all items listed can be administered. [Hazinski, M. (Eds.). (2002). Pediatric advance life support provider manual. American Heart Association. (pp. 156-157).]

4. a

The child's small size does not necessary preclude to thermoregulation problems. A child's ratio of surface area to body mass contributes to rapid heat loss. And the large surface area to body mass allows for more heat loss. Shivering is absent or insufficient in children, especially in those infants younger than 6 months of age. The neuroreceptors for temperature regulation are not mature until approximately 2 years of age. [Holleran, R. (Ed.). (2003). Air and surface patient transport: Principles and practice. (3rd ed.0. St. Louis: Mosby. (pp. 612).]

5. a

Fever may be present, but is not a direct result of the congenital heart disease. Important data to be collected regarding the child with a cyanotic heart defect include: increased cyanosis, vomiting/feeding intolerance, confusion, and stupor. [Krupa, D. (Ed.). (1997). Flight nursing core curriculum. Park Ridge, IL: National Flight Nurses Association. (pp. 545-546).]

6. b

Intussuception can be described as an area of bowel, which telescopes inward upon itself, resulting in a blockage and compression of the intestine. A blockage of the bowel caused by adhesions is described as intestinal atresia. A condition that is typified by the radiologic appearance of pneumotosis intestinalis or free intra-peritoneal air is called necrotizing enterocolitis, and most common in the newborn infant. [McCloskey, K. (Ed.) (1996). Pediatric transport medicine. (1st ed.). St. Louis: Mosby. (pp. 443-444).] [Krupa, D. (Ed.). (1997). Flight nursing core curriculum. Park Ridge, IL: National Flight Nurses Association. (pp. 553).]

7. b

The narrowest portion of the trachea in a child under the age of 8 to 10 is the cricoid cartilage. This structure provides a functional cuff that allows for an appropriate seal without the benefit of the cuff. Using a cuff in smaller children may lead to excessive pressure on the surrounding tissues. [Holleran, R. (Ed.). (2003). Air and surface patient transport: Principles and practice. (3rd ed.). St. Louis: Mosby. (pp. 593).]

8. d

As with any patient, airway and breathing are the first priorities. Hypothermia, fluid volume deficit and decreased cardiac output are important potential problems, but are not a priority over the airway. [Krupa, D. (Ed.). (1997). Flight nursing core curriculum. Park Ridge, IL: National Flight Nurses Association. (pp. 545-546).]

9. a

Blood at the urinary meatus is a sign of urethral injury. Also look for hematoma of the lower abdomen or perineum which is butterfly shaped. Other signs include pallor, increased HR, pain, laceration and bruising. [Slota, M. C. (Ed.) (1998) Core curriculum for pediatric critical care nursing. Philadelphia: Saunders. (pp. 577).]

10. a

Cyanotic heart defects are a result of a lesion causing a right to left shunting of poorly oxygenated or unoxygenated blood back to the peripheral circulation. Ventricular septal defect (VSD) is an acyantoic lesion. [Krupa, D. (Ed.). (1997). Flight nursing core curriculum. Park Ridge, IL: National Flight Nurses Association. (pp. 545-547).]

11. b

The classic "steeple" sign is a significant subglottic narrowing seen on an anterior/ posterior neck film. [Krupa, D. (Ed.). (1997). Flight nursing core curriculum. Park Ridge, IL: National Flight Nurses Association. (pp. 568).]

12. a

Terbutaline is given in a dose of .1mg/kg in 2cc for a nebulizer treatment. To give Terbutaline SQ the dose is .01mg/kg of a 1:1000 solution. Atropine Neb dose is .05-.075mg/kg. Ipratropium bromide is given as a 500-mcg-unit dose. [Holleran, R. (Ed.). (1996). Flight nursing: Principles and practice. (2nd ed.). St. Louis: Mosby. (pp. 506).]

13. a

A serum glucose of less than 40 mg/dl must be treated with an IV bolus of 25% dextrose at 1-2 ml/kg. [Hazinski, M. (Eds.). (2002). Pediatric advance life support provider manual. American Heart Association. (pp. 134).]

14. b

Cranial nerve III is the Oculomotor and its function is pupillary constriction, elevation of upper eye lids, and most of the extraocular movements. [Slota, M. C. (Ed.) (1998) Core curriculum for pediatric critical care nursing. Philadelphia: Saunders. (pp. 301).]

15. c

Incomplete transection causes variable levels of vasomotor instability and bowel and bladder dysfunction. Other symptoms include: asymmetric flaccid paralysis, asymmetric loss of reflexes and varying sensory function below the injury level. [Slota, M. C. (Ed.) (1998) Core curriculum for pediatric critical care nursing. Philadelphia: Saunders. (pp. 347).]

16. c

This child has an incarcerated hernia which is a surgical emergencies requiring immediate repair. ABD surgical emergencies are primarily lesions of obstruction of the GI tract with resulting ileus and gastric distention. Decompression and evacuation of these cavities until further surgical intervention can be accomplished is the goal during transport. [Krupa, D. (Ed.). (1997). Flight nursing core curriculum. Park Ridge, IL: National Flight Nurses Association. (pp. 553-559.]

17. a

The key to treatment of PPHN is early recognition and early interventions. During transport the team wants to reduce elevated pulmonary vascular resistance(PVR). [Slota, M. C. (Ed.) (1998) Core curriculum for pediatric critical care nursing. Philadelphia: Saunders. (pp. 127).]

18. a

High oxygen content and serum alkalosis assist in reducing elevated PVR. Therefore hyperoxia (PaO2 > 100) and hyperventilation (PaCO2 25-35) are the goals of mechanical ventilation therapy. [Slota, M. C. (Ed.) (1998) Core curriculum for pediatric critical care nursing. Philadelphia: Saunders. (pp. 127).]

19. d

If the airway is patent, the crew needs to check a blood glucose level. Hypoglycemia is a complication that can occur while treating DKA. Other complication includes neurological deficits and cerebral edema. [Slota, M. C. (Ed.) (1998) Core curriculum for pediatric critical care nursing. Philadelphia: Saunders. (pp. 416).]

20. b

Hamman's sign is a crunching sound ausculated to the anterior chest that is synchronized to the patient's heartbeat. [Holleran, R. (Ed.). (2003). Air and surface patient transport: Principles and practice. (3rd ed.). St. Louis: Mosby. (pp. 604).]

Environmental Emergencies Transport

1. A patient who sustains snakebite must be transported. Upon arrival at the scene, the flight crew's first priority should be to:

    a. survey the scene before providing patient care.
    b. initiate airway interventions if the patient's airway cannot be maintained.
    c. prepare and administer the appropriate antivenin if the patient is symptomatic.
    d. perform a secondary survey to identify any fang or bite marks on the patient's body.

2. Adequate fluid management of a patient who sustained a burn injury may be demonstrated by:

    a. capillary refill of 3 seconds.
    b. obtaining palpable central pulses only
    c. urinary output of 40 ml/hour in an adult.
    d. urinary output of 2 ml/hour in a child who weighs 30 kg.

3. A patient who has sustained any injury that violates skin integrity is at high-risk for which of the following nursing diagnoses?

    a. Infection from wound contamination
    b. Fluid volume deficit from a bite wound
    c. Ineffective thermal regulation related to the transport environment
    d. Injury related to lack of knowledge about the environment in which he or she has been exposed

4. Which of the following medications administered to an injured patient may put the patient at risk for hypothermia during transport?

    a. Atropine
    b. Cimetidine
    c. Epinephrine
    d. Vecuronium

5. Initial care of a frostbite injury during transport should focus on:

    a. rapid rewarming via warm water soaks to prevent wound infection.
    b. performing a fasciotomy of the injured area to minimize vascular damage.
    c. performing passive range of motion exercises to the injured area to prevent loss of function.
    d. protecting the injured area from trauma or partial thawing, which may lead to further tissue destruction.

6. An 18-year-old man sustains second-degree burns to the anterior aspect of his chest, as well as circumferential burns to his arms and legs. These burns cover approximately what percent of the patient's total body surface area?

    a. 18%
    b. 36%
    c. 72%
    d. 82%

7.   A complication of an electrical burn or the treatment of the burn that may lead to renal failure is:

a.   circumferential injury to the kidneys.
b.   overzealous fluid resuscitation following the injury.
c.   the release of myoglobin from damaged tissue.
d.   electrolyte abnormalities from cellular damage.

8.   A complication frequently associated with submersion emergencies is:

a.   infection.
b.   hypothermia.
c.   myoglobinuria.
d.   metabolic alkalosis.

9.   The first step in the management of a patient with a burn injury is to:

a.   stop the burning process.
b.   provide pain management with narcotics.
c.   provide airway management with nasotracheal intubation
d.   initiate fluid resuscitation with lactated Ringer's solution.

10.   Which of the following would you watch to monitor for rhabdomyolyis in the heat stroke victim?

a.   LOC
b.   Respiratory status
c.   Changes in pulse pressure
d.   Urine output color

11.   Which of the following is a symptom of "after drop" in the hypothermic patient?

a.   Drop in heart rate and shivering
b.   Normothermia and hypoventilation
c.   Hyperthermia and hypertension
d.   Cardiac dysrhythmias and hypotension

12.   You arrive at a small clinic for a 20-year old female who has been on a diving trip. She is complaining of headache, nausea, and abdominal and back pain. The transport crew is aware she is exhibiting symptoms of:

a.   Type-1 decompression sickness.
b.   Type-2 decompression sickness.
c.   Type-3 decompression sickness.
d.   Type-4 decompression sickness.

13.   The transport crew knows that water is denser than air. Using this knowledge they know a scuba diver at 33 feet will experience an ambient pressure of:

a.   1 ATM.
b.   2 ATM.
c.   3 ATM.
d.   None of the above.

14. The transport crew recognizes that the first sign of hypothermia is:

    a. loss of shivering.
    b. paradoxical undressing.
    c. superficial frost bite on face & nose.
    d. pale skin.

15. You are preparing to transport a 36-year-old male who was burned in a house fire. The injury occurred three hours ago. The patient has 30% TBSA burned. He weighs 75 kg. He has received 800cc of fluid prior to your arrival. Calculate how much fluid to give over the next 5 hours using the Parkland formula.

    a. 4500 cc
    b. 8650 cc
    c. 3700 cc
    d. 740 cc

16 Which of the following environmental stressors does not affect thermoregulation?

    a. Wind
    b. Moisture
    c. Humidity
    d. Hydration

17. Upon arrival to a scene call you find your patient is a 68 year-old homeless individual in cardiac arrest. The patient has a core temperature of 29°C. Upon assessing this patient which of the following should be done?

    a. Stop resuscitation
    b. Run resuscitation at scene until patient stable
    c. Transport patient regardless of cardiopulmonary status
    d. Call medical command and discuss options

18. You are preparing to transport a patient whose BP is 80/44 and diagnosed with heat stroke. You know that the field IV fluid of choice is:

    a. LR.
    b. NS.
    c. D5 ¼.
    d. D5LR.

19. In preparing to transport the hypothermic patient the transport team recognizes that cerebral blood flow decreases 6-7% for every 1 degree C decline until the temperature reaches:

    a. 30° C
    b. 27° C
    c. 25° C
    d. 21° C

Environmental Emergencies Transport Answers

1. a
The flight nurse should survey the scene at the site of an environmental emergency before entering the patient area. Many potential hazards exist at these scenes that may cause injuries to the flight team, cause aircraft damage, or place rescuers at risk. These need to be identified and managed. [Krupa, D. (Ed.). (1997). Flight nursing core curriculum . Park Ridge, IL: National Flight Nurses Association. (pp. 633).]

2. c
Adequate fluid resuscitation is indicated by blood pressure within normal limits for the patient's age and medical history, palpable peripheral and central pulses, urinary output greater than 30 cc/hr for adults and 1-2 ml/kg/ hour for children. [Krupa, D. (Ed.). (1997). Flight nursing core curriculum. Park Ridge, IL: National Flight Nurses Association. (pp. 637).]

3. a
A patient with any injury that violates skin integrity is at high risk for infection from wound contamination. [Krupa, D. (Ed.). (1997). Flight nursing core curriculum. Park Ridge, IL: National Flight Nurses Association. (pp. 636).]

4. d
Neuromuscular blocking agents such as Vecuronium interfere with the patient's ability to shiver, which is one mechanism of preserving heat. [Holleran, R. (Ed.). (2003). Air and surface patient transport: Principles and practice. (3rd ed.). St. Louis: Mosby. (pp. 469).]

5. d.
During transport, the flight nurse needs to protect the injured part from additional trauma and partial thawing. Protection may include covering the area with a bulky or "tented" dressing and allowing it to stay frozen if there is a possibility that it may refreeze. [Holleran, R. (Ed.). (2003). Air and surface patient transport: Principles and practice. (3rd ed.). St. Louis: Mosby. (pp. 473).]

6. c
Using the Rule of Nines, this patient's injury can be approximated at 72 % of the total body surface area. The anterior chest is 18 %; each arm 9 %, and each leg 18 % (18 + 9 + 9 + 18 + 18 = 72). [Holleran, R. (Ed.). (2003). Air and surface patient transport: Principles and practice. (3rd ed.). St. Louis: Mosby. (pp. 322).]

7. c.
Myoglobin is released from damaged tissue when extensive injury, such as electrical burns, occurs and can accumulate in the renal tubules and cause renal failure. Electrolyte abnormalities are more likely to cause cardiac problems than renal problems. [Holleran, R. (Ed.). (2003). Air and surface patient transport: Principles and practice.(3rd ed.). St. Louis: Mosby. (pp. 334).]

8. b
Immersion hypothermia occurs more rapidly than in non-immersion hypothermia, and heat loss may be greater if the patient swims or treads water. [Holleran, R. (Ed.). (2003). Air and surface patient transport: Principles and practice. (3rd ed.). St. Louis: Mosby. (pp. 474).]

9. a
The first step in the management of the patient who has sustained a burn injury is to stop the burning process. [Krupa, D. (Ed.). (1997). Flight nursing core curriculum. Park Ridge, IL: National Flight Nurses Association. (pp.647).]

10. d
Blood or dark colored urine, muscle cramps, & hyperkalemia are signs of rhabdomyolyis. [Holleran, R. (Ed.). (2003). Air and surface patient transport: Principles and practice. (3rd ed.). St. Louis: Mosby. (pp. 487).]

11. d
After drop occurs when rewarming a patient. As the periphery is warmed acidotic blood is dumped into the central circulation causing a drop in BP and dysrhythmias to occur. [Holleran, R. (Ed.). (2003). Air and surface patient transport: Principles and practice. (3rd ed.). St. Louis: Mosby. (pp. 475).]

12. b
Type-II decompression sickness is more serious than Type-I. Typical symptoms include sensory and visual disturbances, SOB, paresthesia, fatigue, weakness, H/A, nausea, chest pain, ABD pain, back pain, bowel and bladder dysfunction as well as LOC. Type I symptoms include H/A, fatigue, limb or joint pain, skin rash and localized swelling. Type 4 is also known, as "chokes" is pulmonary decompression illness. Signs include SOB, chest pain and nonproductive cough. [Holleran, R. (Ed.). (2003). Air and surface patient transport: Principles and practice. (3rd ed.). St. Louis: Mosby. (pp. 500).]

13. b
Because the density of water is uniform throughout, the proportional relationship of pressure and depth remains constant: pressure increases 1 ATM for every 33 feet (10-m) column seawater. At sea level a column of seawater is 1 ATM. [Holleran, R. (Ed.). (2003). Air and surface patient transport: Principles and practice. (3rd ed.). St. Louis: Mosby. (pp. 497).]

14. b
Paradoxical undressing is the first sign of severe hypothermia. . [Krupa, D. (Ed.). (1997). Flight nursing core curriculum. Park Ridge, IL: National Flight Nurses Association. (pp.643).]

15. c
To calculate using the parkland formula: 4 cc X wt in kg X % TBSA burned, then 4cc X 75kg X 30% = 9000cc. You give ½ of that dose in the first 8 hours from time the burn occurred and give the other ½ over the next 16 hours. The patient needs 4500cc in the first 8 hours. He has already received 800cc, so he needs 3700cc over the next 5 hours. [Holleran, R. (Ed.). (1996). Flight nursing: Principles and practice. (2nd ed.). St. Louis: Mosby. (pp. 296).]

16. d
Hydration may affect thermoregulation, but it is not an environmental stressor. . [Krupa, D. (Ed.). (1997). Flight nursing core curriculum. Park Ridge, IL: National Flight Nurses Association. (pp. 638).]

17. c
All hypothermic patients must be transported regardless of cardiopulmonary status. A patient is not dead, until they are warm and dead. This is defined as 32° Celsius. [Holleran, R. (Ed.). (2003). Air and surface patient transport: Principles and practice. (3rd ed.). St. Louis: Mosby. (pp. 475).]

18. b
Because of complications of impaired cardiac function, pulmonary catheters best guide pulmonary edema, CHF, ARDS, and kidney function fluid resuscitation. Field fluid replacement is NS until SBP is 90 or better. [Holleran, R. (Ed.). (2003). Air and surface patient transport: Principles and practice. (3rd ed.). St. Louis: Mosby. (pp. 491).]

19. c
Cerebral flow decreases 6-7% for every 1° C decline until 25° C is reached.
[Holleran, R. (Ed.). (2003). Air and surface patient transport: Principles and practice. (3rd ed.). St. Louis: Mosby. (pp. 470).]

Patient Management in Transport

1. The following blood gas result indicates what respiratory or metabolic problem? PaO2 88, pCO2 60, Bicarb 22, pH 7.32

    a. Respiratory acidosis
    b. Metabolic acidosis
    c. Respiratory alkalosis
    d. Metabolic alkalosis

2. Upon arrival to transport a head injury patient currently being ventilated with an ambu bag via an ETT, the transport team has the following blood gas result to interpret: paO2 235, pCO2 18, Bicarb 20, pH 7.60

    a. Respiratory acidosis
    b. Metabolic acidosis
    c. Respiratory alkalosis
    d. Metabolic alkalosis

3. Appropriate care of the patient in question #2 would include which of the actions?

    a. Increasing tidal volume
    b. Decreasing ventilation rate
    c. Administering bicarbonate
    d. Adding PEEP

4. The transport team is called to transport a 14-year-old patient to a pediatric intensive care unit. On arrival the patient's respiratory rate is 28 with deep inspiration and expiration. PaO2 158, pCO2 15, Bicarb 8, pH 7.05. The blood gas result indicates what acid base disturbance.

    a. Respiratory alkalosis and metabolic alkalosis
    b. Respiratory acidosis and metabolic acidosis
    c. Respiratory acidosis and metabolic acidosis
    d. Respiratory alkalosis and metabolic acidosis

5. A 12-lead EKG shows a 3mm ST elevation in leads II, III, and aVF. This is indicative of an ischemic pattern for which area of the myocardium?

    a. Inferior wall
    b. Anterior wall
    c. Septum
    d. Lateral wall

6. During the transport of a 68 year old male for unstable angina he begins to complain of increasing chest pain. The 12 lead done during this episode of chest pain has ST elevation of V1 and V2 of 5mm. This is indicative of ischemia of which portion of the myocardium?

    a. Inferior wall
    b. Anterior wall
    c. Septum
    d. Lateral wall

7.   An anterolateral ischemic pattern would be seen in what leads?

      a.   II, III, aVF
      b.   AVL, V1, V2
      c.   I, aVL, II, aVF
      d.   I, V3, V4, V5, V6

8.   Central venous pressure monitors:

      a.   intra-arterial pressure.
      b.   pulmonary artery pressure.
      c.   right atrial pressure.
      d.   femoral venous pressure.

9.   Central venous pressure (CVP) monitoring can assist in all the following except:

      a.   measure cardiac output.
      b.   guide for fluid replacement.
      c.   administer blood products, TPN or drug therapy.
      d.   obtain central venous blood samples.

10.   A pulmonary artery pressure monitoring system is reflecting the filling pressure of:

      a.   right atrium.
      b.   left atrium.
      c.   right ventricle.
      d.   left ventricle.

11.   A normal pulmonary capillary wedge pressure (PCWP) reading is:

      a.   1 – 3 mm Hg
      b.   6 – 12 mm Hg
      c.   15 – 18 mm Hg
      d.   20 – 25 mm Hg

12.   The central line readings obtained in the ICU prior to transport of a patient are as follows: CVP 13, Cardiac index (CI) 1.4, and a PCWP 18. This could indicate what problem for the patient?

      a.   Hypovolemia
      b.   Heart failure
      c.   Anaphylactic shock
      d.   None of the above

13.   The intra-aortic balloon pump (IABP) is indicated in all the following patients except:

      a.   Cardiogenic shock
      b.   Weaning from cardiopulmonary bypass
      c.   Unstable angina
      d.   Aortic insufficiency

14. The balloon of the IABP is inflated throughout:

    a. Cardiac cycle
    b. Diastole
    c. Systole
    d. Unrelated to the cardiac cycle

15. While monitoring the arterial line of a patient with an IABP, the balloon should show diastolic augmentation at what part of the arterial waveform?

    a. systolic phase
    b. ventricular ejection phase
    c. dicrotic notch
    d. None of the above

16. Complications of the IAPB include which of the following?

    a. Emboli
    b. Thrombocytopenia
    c. Rupture of the balloon
    d. All of the above

17. During transport while monitoring the balloon pump waveforms a rounded waveform is noted. This type of wave form is frequently due to:

    a. Gas loss
    b. Sustained inflation
    c. Catheter kink
    d. None of the above

18. A ventricular assist device (VAD) is indicated in which of the following patients?

    a. Acute MI in cardiogenic shock
    b. Postcardiotomy ventricular failure who cannot be weaned from cardiopulmonary bypass
    c. Candidates for cardiac transplantation whose conditions deteriorate prior to a donor
    d. All of the above

19. A VAD allows the ventricle to rest by:

    a. providing increased blood flow to the coronary arteries.
    b. enhancing left ventricular emptying.
    c. diverting blood from the natural ventricle to an artificial pump.
    d. relieving pulmonary congestion.

20. Synchronous Intermittent mandatory ventilation (SIMV) is described as:

   a. ventilator deliver breaths at a preset interval with spontaneous breathing allowed between ventilator-administered breaths.
   b. ventilator delivers preset breaths in coordination with the respiratory effort of the patient. Spontaneous breathing is allowed between breaths.
   c. ventilator delivers preset breaths in coordination with the respiratory effort of the patient. With each inspiratory effort, the ventilator delivers a full assisted tidal volume.
   d. breaths are delivered at preset intervals, regardless of patient effort.

21. During transport of a patient you will be using a ventilator. You can calculate the initial tidal volume using:

   a. 5 - 9 ml/kg.
   b. 10 – 15 ml/kg.
   c. 16 – 20 ml/kg.
   d. 21 - 25 ml/kg.

22. A peak inflation pressure of _____ is associated with an increase in barotraumas.

   a. 10 cm H2O
   b. 20 cm H2O
   c. 30 cm H2O
   d. >40 cm H2O

23. You are in transit to receive a multiple trauma patient. You receive blood gas results, which of the following blood gas results would have you preparing to intubate and ventilate the patient?

   a. PaO2 88, pCO2 44, Bicarb 20, pH 7.38
   b. PaO2 80, pCO2 48, Bicarb 26, pH 7.35
   c. PaO2 50, pCO2 75, Bicarb 16, pH 7.03
   d. PaO2 158, pCO2 15, Bicarb 8, pH 7.05

24. On arrival at a facility a patient has just been intubated. The patient is a 48 year-old-male with severe head injuries and no known lung disease. The patient weighs 70kg. Post intubation pulse oximetry readings are 96 – 98 and end tidal CO2 is 35 – 38. What would your initial ventilator settings be?

   a. O2:100% TV:560cc Rate:12 I:E ratio:1:2 PEEP: 3
   b. O2:100% TV:750cc Rate:6 I:E ratio:1:2 PEEP: 3
   c. O2:100% TV:560cc Rate:12 I:E ratio:1:4 PEEP: 10
   d. O2:100% TV:1000cc Rate:6 I:E ratio:1:4 PEEP: 10

25. To control the CO2 while on a ventilator, you manipulate the:

   a. respiratory rate and FiO2.
   b. FiO2 and PEEP.
   c. PEEP and tidal volume.
   d. respiratory rate and tidal volume.

Patient Management in Transport Answers

1. a

Respiratory acidosis. The paO2 indicates hypoxemia while the elevated pCO2 indicates a respiratory acidosis. The bicarbonate level of 22 is within normal parameters indicating no metabolic component at this time. The pH confirms the acidosis. [Krupa, D. (Ed.). (1997). Flight nursing core curriculum. Park Ridge, IL: National Flight Nurses Association. (pp. 107, 114).] [Marino, P. (1998). The ICU handbook. (2nd ed.). Baltimore: Williams & Wilkins. (pp. 583-590)].

2. c

Respiratory alkalosis. The paO2 indicates the patient is being well oxygenated, with increased ventilation causing a decrease in the pCO2 creating a respiratory alkalosis. The bicarbonate level is normal at this time indicating no metabolic component. The pH confirms the alkalosis. [Krupa, D. (Ed.). (1997). Flight nursing core curriculum. Park Ridge, IL: National Flight Nurses Association. (pp. 109).] [Marino, P. (1998). The ICU handbook. (2nd ed.). Baltimore: Williams & Wilkins. (pp. 583-590)]:

3. b

Decreasing respiratory rate. Based on the above blood gases the patient is being ventilated. It would be acceptable to decrease tidal volume as well or in combination with decreasing the rate to allow the CO2 and pH to return to near normal. Administering bicarbonate does not treat a respiratory alkalosis and administering PEEP is not indicated with a pCO2 of 18. [Krupa, D. (Ed.). (1997). Flight nursing core curriculum. Park Ridge, IL: National Flight Nurses Association. (pp. 110).] [Marino, P. (1998). The ICU handbook. (2nd.ed.). Baltimore: Williams & Wilkins. (pp 583-590)].

4. d

pCO2 is decreased indicating a respiratory alkalosis; the bicarbonate level is low indicating metabolic acidosis. The pH confirms a significant acidosis. [Krupa, D. (Ed.). (1997). Flight nursing core curriculum. Park Ridge, IL: National Flight Nurses Association. (pp. 109).] [Marino, P. (1998). The ICU handbook. (2nd. ed.). Baltimore: Williams & Wilkins. (pp. 583-590).]

5. a

The inferior wall of the myocardium is indicated by leads II, III, and aVF. [Holleran, R. (Ed.). (2003). Air and surface patient transport: Principles and practice. (3rd ed.). St. Louis: Mosby. (pp. 367).]

6. c

Leads V1 and V2 look primarily at the septum of the heart. [Holleran, R. (Ed.). (2003). Air and surface patient transport: Principles and practice. (3rd ed.). St. Louis: Mosby. (pp. 367).]

7. d

Leads I, aVL, V5 and V6 are reflective of the lateral wall and two contiguous leads are needed to indicate a pattern, therefore I, V5 and V6 indicate the lateral wall problem. V2, V3, V4, and V5 are indicative of an anterior wall problem, so V3, 4, and 5 lead to the anterior pattern interpretation. When the two walls are both indicating a pattern there is an anterolateral pattern. [Holleran, R. (Ed.). (2003). Air and surface patient transport: Principles and practice. (3rd ed.) St. Louis: Mosby. (pp. 367).]

8. c

A central venous pressure-monitoring device measures the right atrial pressure or the pressure of the great veins within the thorax. A CVP reading best reflects the right side of the cardiac function. [Holleran, R. (Ed.). (2003). Air and surface patient transport: Principles and practice. (3rd ed.). St. Louis: Mosby. (pp. 401-402).]

9. a
A central venous catheter can be used as a guide for fluid replacement, monitor pressures in the right atrium and central veins, administer blood products, TPN and drug therapy, obtain venous access when peripheral vein sites are inadequate, insert a temporary pacemaker and obtain venous blood samples. It does not provide numbers needed to measure cardiac output. [Holleran, R. (Ed.). (2003). Air and surface patient transport: Principles and practice. (3rd ed.). St. Louis: Mosby. (pp. 401-402).]

10. d
A pulmonary artery catheter is designed to reflect the left ventricular end diastolic pressure – the filling pressure of the left ventricle. [Holleran, R. (Ed.). (2003). Air and surface patient transport: Principles and practice. (3rd ed.). St. Louis: Mosby. (pp. 401-402).]

11. b
Normal pulmonary capillary wedge pressure is 6 – 12 mmHg. [Holleran, R. (Ed.). (2003). Air and surface patient transport: Principles and practice. (3rd ed.). St. Louis: Mosby. (pp. 402).]

12. b
Heart failure is indicated by an elevated CVP 13 (normal 1-6mm Hg), a low cardiac index 1.4 (normal 2.4 – 4.0 L/min/m2), and elevated PCWP 18, (normal 6 – 12 mmHg). A low CVP, low CI, and high PCWP indicate Hypovolemia, and anaphylaxis would be indicated with readings of a low CVP, high CI, and low PCWP. [Holleran, R. (Ed.). (2003). Air and surface patient transport: Principles and practice. (3rd ed.). St. Louis: Mosby. (pp. 402).]

13. d
Intra-aortic balloon pumps are contraindicated in patients with aortic insufficiency, dissecting aortic aneurysm, chronic end-stage heart disease, and severe peripheral vascular disease. Common indications for the use of the IABP include: cardiogenic shock, awaiting cardiopulmonary bypass, weaning off cardiopulmonary bypass, papillary muscle rupture or ventricular septal defect post infarction, unstable angina resistant to medical therapy, and as a bridge to cardiac transplantation. [Holleran, R. (Ed.). (2003). Air and surface patient transport: Principles and practice. (3rd ed.). St. Louis: Mosby. (pp. 379-380).]

14. b
The balloon is inflated throughout diastole – the period during the cardiac cycle when blood is not being pumped forward by the heart. [Holleran, R. (Ed.). (2002). Air and Surface Patient Transport: Principles and Practice. 3rd ed. St. Louis: Mosby. (pp. 380).]

15 c
The dicrotic notch is the place in the arterial waveform that shows the beginning of augmented diastole as the balloon inflates. [Quaal, S.J., (1993). Comprehensive intra-aortic balloon pumping. (2nd ed.). St. Louis: Mosby. (pp. 132-133).]

16. d
Several complications are associated with the use of the IABP and include emboli, thrombosis, thrombocytopenia, infection, rupture of the aorta, rupture of the balloon, impaired circulation, bleeding, and inability to wean. [Holleran, R. (Ed.). (2003). Air and surface patient transport: Principles and practice. (3rd ed.) St. Louis: Mosby. (pp. 379-380).]

17. c
A rounded balloon waveform may be caused by a kink in the catheter tubing, improper IAB catheter position, sheath not being pulled back to allow inflation of the IAB, the IAB is too large for the aorta, not fully unwrapped or there is H2O condensation in the external tubing. [Datascope Clinical Support Services. (1998). Advanced seminar for intra-aortic balloon pumping. Fairfield, N.J.]

18. d
All of these patients would be candidates for a ventricular assist device. [Holleran, R. (Ed.). (2003). Air and surface patient transport: Principles and practice. (3rd ed.) St. Louis: Mosby. (pp. 380).]

19. c
The VAD allows the myocardium to rest by diverting blood from the natural ventricle to an artificial pump that maintains the circulation. [Holleran, R. (Ed.). (2003). Air and surface patient transport: Principles and practice. (3rd ed.) St. Louis: Mosby. (pp. 380).]

20. b
SIMV (synchronous intermittent mandatory ventilation)- The ventilator delivers preset breaths in coordination with the respiratory effort of the patient. Spontaneous breathing is allowed between breaths. Synchronization attempts to limit the barotrauma, which may occur with IMV when a preset breath is delivered to a patient who is already maximally inhaled (breath stacking) or is forcefully exhaling. IMV (intermittent mandatory ventilation) breaths are delivered at a preset interval, and spontaneous breathing is allowed between ventilator-administered breaths. Spontaneous breathing occurs against the resistance of the airway tubing and ventilator valves, which may be formidable. A/C (Assist-control ventilation) delivers preset breaths in coordination with the respiratory effort of the patient. With each inspiratory effort, the ventilator delivers a full assisted tidal volume. Spontaneous breathing independent of the ventilator between A/C breaths is not allowed. CMV (continuous mandatory ventilation) breaths are delivered at preset intervals, regardless of patient effort. [Joyce, D. (2001). Ventilator Management. www.emedicine.com]

21. b
Calculation of tidal volume begins with 10 – 15 ml/kg. [Holleran, R. (Ed.). (2003). Air and surface patient transport: Principles and practice. (3rd ed.) St. Louis: Mosby. (pp. 419).]

22. d
A peak inflation pressure of > 40 cm H_2O is associated with an increase in barotraumas. [Joyce, D. (2001). Ventilator Management. www.emedicine.com]

23. c
PaO_2 50, pCO_2 75, Bicarb 16, pH 7.03 – Laboratory criteria for intubation and mechanical ventilation include PaO_2 <55 mm Hg, $PaCO_2$ >50 mm Hg, and pH <7.32. [Joyce, D. (2001). Ventilator Management. www.emedicine.com]

24. a
O_2:100% TV:560cc Rate:12 I:E ratio:1:2 PEEP: 3 Observations of the adverse effects of barotrauma and volutrauma have led to recommendations of tidal volumes of 5.0-10 mL/kg. The goal is to adjust the TV so that plateau pressures are less than 35 cm H_2O. A respiratory rate (RR) of 8-12 breaths per minute is recommended. High rates allow less time for exhalation, increase mean airway pressure, and cause air trapping. The normal inspiration/expiration (I/E) ratio to start is 1:2. It is common to apply physiologic PEEP of 3.0-5.0 cm H_2O to prevent decreases in functional residual capacity in those with normal lungs. [Joyce, D. (2001). Ventilator Management. www.emedicine.com]

25. d
To control the pH and pCO_2 you manipulate the minute ventilation, i.e. the respiratory rate and tidal volume. To control oxygenation you adjust the FiO_2 and the mean airway pressure (PEEP). [Tegtmeyer, K. (1998). Initial Mechanical Ventilation. www.peds.umn.edu]

Aviation, Safety, Management, Education, Program Management, and Professionalism

1. According to federal aviation regulations, responsibility for all aspects of the safe operation of the aircraft rests with:

    a. the pilot in command.
    b. the program safety officer.
    c. the program medical director.
    d. air medical team members on board the aircraft and the pilot in command.

2. A "sterile cockpit" should occur:

    a. during engine starts.
    b. on approaches and landings.
    c. while flying in air traffic congested areas.
    d. in all of the above situations.

3. All of the following are crash positions except:

    a. with shoulder harness on, bent over with arms folded across lower legs.
    b. with shoulder harness on, sitting upright with the back of the head against the headrest and the feet flat on the floor.
    c. lap belt only and forward facing, bent over with arms folded across lower legs.
    d. lap belt only and side sitting, sitting upright with the back of the head against the headrest and the feet flat on the floor.

4. When exiting the aircraft after an emergency, the flight team should initially meet at which position off the nose?

    a. 12 o'clock
    b. 3 o'clock
    c. 6 o'clock
    d. 9 o'clock

5. The emergency locator transmitter (ELT) frequency is heard on:

    a. 120.5 VHF.
    b. 121.5VHF.
    c. 141.5 VHF.
    d. 212.5 VHF.

6. The air medical team's first priority when a pilot announces an aircraft emergency is to:

    a. calm the patient.
    b. assume crash positions.
    c. secure loose equipment.
    d. transmit the aircraft's position to flight following.

7. What is the proper "crash position" to assume in preparation for a hard landing when wearing a four-point restraint?

   a. Lie flat on the stretcher
   b. Grasp your knees and bend forward
   c. Sit upright in your seat and tighten all belts
   d. Loosen your seat belt and stretch out as much as possible

8. Seat belts must be worn during:

   a. all phases of flight.
   b. take-off and landing.
   c. only after the pilot declares an emergency.
   d. only when they do not interfere with patient care.

9. Which of the following regulatory agencies may have authority related to aspects of the air medical program?

   a. Federal Aviation Administration (FAA).
   b. Federal Communications Commission (FCC).
   c. Health Care Financing Agency (HCFA).
   d. All of the above agencies.

10. Which of the following elements is not typically included in the air medical transport service's marketing plan?

    a. Financial plan
    b. Market research and analysis
    c. Measurable marketing objectives
    d. Mission statement and program goals

11. In the air medical transport service manager's role to help optimize reimbursement for transport, the manager should perform all of the following functions except:

    a. work with the Medicare intermediary to appeal denials.
    b. track payor mix and the amount (percentage) collected from each payor.
    c. work with the managed care organizations to develop transport contracts.
    d. pursue only those contracts that reimburse air medical transport services more than 75%.

12. Most emergency medical technicians (EMTs) function in the field under the license of the:

    a. regional EMS coordinator.
    b. ambulance service director.
    c. state, which grants EMTs individual licenses.
    d. emergency medical service (EMS) medical director.

13. The initial intervention (within hours) after a critical incident is:

    a. defusing.
    b. debriefing.
    c. peer counseling.
    d. professional counseling.

14. Critical incident stress is defined as:

    a. an "emotional aftershock" to an incident.
    b. an exposure to intense, prolonged stress.
    c. a state of physical and psychological arousal.
    d. many stresses experienced in a short period of time.

15. What occurs physiologically when the "fight-or-flight" response is activated?

    a. Heart rate decreases
    b. Respirations increase
    c. Pulse pressure narrows
    d. Blood glucose levels drop

16. Which of the following radio bands are not specifically designated as a public service frequency?

    a. UHF 3O to 5O MHz
    b. UHF 450 to 470 MHz
    c. VHF low-band FM 30 to 50 MHz
    d. VHF high-band FM 148 to 174 MHz

17. From an air medical communications perspective, the activities during the transport phase of a scene response include:

    a. information about the nature of the call.
    b. determining the accepting physician's name.
    c. obtaining detailed information about the landing zone.
    d. taking patient report and establishing on-line medical control if necessary.

18. When transporting patients outside the United States, the air medical team need not be concerned about:

    a. any civil unrest in the area.
    b. the location of the American Embassy.
    c. formulating a plan if team members are separated.
    d. altitude restrictions for the air space in that country.

19. All of the following situations require emergency extrication of the patient except:

    a. possible explosion.
    b. potential collapse of structure.
    c. deterioration of patient condition.
    d. entrapment in a vehicle consumed in flames.

20. Of the following, which list presents the stages of a rescue from a motor vehicle accident from greatest to least in priority?

    a. Assessing and controlling hazards, gaining access, entering the vehicle, then providing urgent patient care
    b. Assessing and controlling hazards, gaining access, providing urgent patient care, then entering the vehicle
    c. Gaining access, assessing and controlling hazards, entering the vehicle, then providing urgent patient care
    d. Providing urgent patient care, assessing and controlling hazards, gaining access, then entering the vehicle

21. Conditions found in the primary survey which require immediate patient extrication include:

    a. cardiac arrest.
    b. respiratory arrest.
    c. bleeding or shock that cannot be controlled.
    d. any of the above conditions.

22. What is the most important step in disaster preparedness?

    a. Conducting practice drills
    b. Ensuring that a disaster plan is in place
    c. Ensuring that there is adequate personnel available
    d. Ensuring that there is appropriate equipment available

23. Which of the following is not considered a primary role of the air medical team in a disaster?

    a. Acting as medical control
    b. Providing patient care at the scene
    c. Providing patient care during transport
    d. Transporting supplies and medical teams

24. One way to manage aircraft resources at a disaster is to identify:

    a. a staging area.
    b. a landing zone.
    c. an incident commander.
    d. a dedicated frequency for radio communications.

25. Disaster planning at the local, regional, or state level should focus on:

    a. getting more help.
    b. preparing for all types of disasters.
    c. preparing for large-scale disasters only.
    d. preparing for the type of disaster that is most likely to occur in the area.

26. Which of the following categories is not an accepted way to classify patients during triage?

    a.   Emergent
    b.   Urgent, hospital care needed
    c.   Non-urgent, but hospital care needed
    d.   Walking wounded, no hospital treatment needed

27. In the assessment phase of the teaching process, it is important to:

    a.   develop specific goals.
    b.   determine the learner's strengths.
    c.   provide feedback on performance.
    d.   assist the learner in achieving goals.

28. During the orientation period, a new flight nurse has repeated errors in multiple areas. The preceptor should first:

    a.   identify the nurse's barriers to learning.
    b.   reassign the nurse to another preceptor.
    c.   initiate stress reduction techniques to facilitate learning.
    d.   demonstrate proper use of all equipment and procedures to reinforce the nurse's learning base.

29. Outreach education methods include which of the following?

    a.   Doing the best job possible on every flight
    b.   Presenting a slide lecture on landing zone safety
    c.   Providing educational sessions to referring agencies
    d.   All the above

30. When providing education to a pediatric patient, the flight nurse must remember that:

    a.   the child has an independent personality.
    b.   motivation occurs through internal rewards.
    c.   the child's experience is to be used as a resource.
    d.   readiness to learn depends on age and developmental level.

31. Flight nursing practice requires that nurses:

    a.   practice in a rotor wing aircraft only.
    b.   implement a safe transport environment.
    c.   need only limited physical and emotional preparation.
    d.   function independently from other flight team members.

32. Implementation involves:

    a.   obtaining and documenting patient data.
    b.   providing needed patient care interventions.
    c.   identifying individual patient care outcomes.
    d.   evaluation of the effects of administered medications.

33. The Standards of Flight Nursing Practice are best described as:

    a. law.
    b. rules.
    c. guidelines.
    d. regulations.

34. Of the following, who is responsible for clearly documenting the medical necessity for transfer of a patient?

    a. Flight nurse
    b. Referring physician
    c. Receiving physician
    d. Program medical director

35. Which of the following general principles of law is based on vicarious liability?

    a. Indemnification
    b. Intentional torts
    c. Respondeat superior
    d. Good Samaritan laws

36. Ethical guidelines for informed consent do not include the right to:

    a. privacy.
    b. anonymity.
    c. self-determination.
    d. appropriate remuneration.

37. The dependent variable can be distinguished from the independent variable in that:

    a. there is no relationship between the two variables.
    b. the dependent variable is the object being manipulated and the independent variable is the object being studied.
    c. the dependent variable is the object being studied and the independent variable is the object being manipulated.
    d. both variables are manipulated by the researcher at different points in the research process.

38. What type of nursing research is most often reported in the literature?

    a. Applied
    b. Qualitative
    c. Fundamental
    d. Quantitative

Aviation, Safety, Management, Education, Program Management, and Professionalism Answers

1. a
Although all air medical crew members should be alert for safety, the pilot in command has responsibility according to the Federal Aviation Regulations. [Krupa, D. (Ed.). (1997). Flight nursing core curriculum. Park Ridge, IL: National Flight Nurses Association. (pp. 674).]

2. d
A "sterile cockpit" is the time in which there should be no unnecessary communication. Maintaining a sterile cockpit is important during any time when emergency procedures are most likely to occur, including during engine starts, approaches and landings, and when flying in any congested air traffic areas. [Krupa, D. (Ed.). (1997). Flight nursing core curriculum. Park Ridge, IL: National Flight Nurses Association. (pp. 678).]

3. a
Shoulder harnesses allow the crash position to be upright. When sitting sideways with or without a shoulder harness however the correct position is upright. And without a shoulder harness and sitting facing forward the 'classic' bent over position is the appropriate one to be in. [Krupa, D. (Ed.). (1997). Flight nursing core curriculum. Park Ridge, IL: National Flight Nurses Association. (pp. 682).]

4. a
The 12 o'clock position is the preferred meeting place after an emergency exit, if it is available. [Krupa, D. (Ed.). (1997). Flight nursing core curriculum. Park Ridge, IL: National Flight Nurses Association. (pp. 686).]

5. b
The frequency of the emergency locator transmitter is 121.5. [Krupa, D. (Ed.). (1997). Flight nursing core curriculum. Park Ridge, IL: National Flight Nurses Association. (pp. 683).]

6. c
Not all air medical programs have medical crew access to the radio. While communication is the first priority, this is usually accomplished by the pilot. Therefore, it is the team's first priority to secure loose equipment. [Krupa, D. (Ed.). (1997). Flight nursing core curriculum. Park Ridge, IL: National Flight Nurses Association. (pp. 702-703).]

7. c
The correct hard landing position with a four point restraint is to secure the belt and sit upright securing arms under the shoulder harness holding onto the shoulder strap. Leaning forward with the head placed between the knees and encircling the knees and head with the forearms is the proper position when using a lap belt only. Loosening the seat belt will not secure the crew member in the event of a crash. [Krupa, D. (Ed.). (1997). Flight nursing core curriculum. Park Ridge, IL: National Flight Nurses Association. (pp. 703).]

8. a
Seat belts should be worn during all phases of flight; but, they must be worn for take-off and landing. A safety harness should be used when unbelted to deliver patient care during flight. [Krupa, D. (Ed.). (1997). Flight nursing core curriculum. Park Ridge, IL: National Flight Nurses Association. (pp. 701).]

9. d
Air medical programs follow the guidelines of many regulatory agencies; some programs are regulated by state agencies as well as the agencies listed. The FAA regulates all aviation aspects of an air medical program. The FCC regulates all radio and telephone frequencies and communications. HCFA regulates the reimbursement of Medicare and Medicaid. There are similar agencies in other countries as well. For example, Transport Canada legislates the aviation aspect of air medical transport in Canada. [Krupa, D. (Ed.). (1997). Flight nursing core curriculum. Park Ridge, IL: National Flight Nurses Association. (pp. 724).]

10. a
A financial plan is included in the business plan. The marketing plan should include volume goals and measurable objectives based on the mission of the program. Market research and analysis should drive the marketing plan. [Krupa, D. (Ed.). (1997). Flight nursing core curriculum. Park Ridge, IL: National Flight Nurses Association. (pp. 722-723).]

11. d
One part of the service manager's role is to optimize reimbursement. Working with Medicare intermediaries, tracking payor mix, and obtaining contracts with managed care organizations could certainly assist in this goal. Pursuing only contracts that reimburse at a rate of 75%, although certainly advantageous would not keep a program transporting since so many insurances are reducing their reimbursement rates. [Krupa, D. (Ed.). (1997). Flight nursing core curriculum. Park Ridge, IL: National Flight Nurses Association. (pp. 721-722).]

12. d
While Emergency Medical Technicians (EMT's) need individual certification, they function in the field under the medical license of the emergency medical service director. [Krupa, D. (Ed.). (1997). Flight nursing core curriculum. Park Ridge, IL: National Flight Nurses Association. (pp. 738).]

13. c
The initial phase of critical incident stress debriefing is called defusing. Debriefing occurs 24 to 72 hours after the event. Peer counseling or professional counseling are generally separate from defusing and debriefing. [Krupa, D. (Ed.). (1997). Flight nursing core curriculum. Park Ridge, IL: National Flight Nurses Association. (pp. 752-753)].

14. a
Critical incident stress is often referred to as an "emotional aftershock" to an incident. The usual ability to cope and function is overwhelmed. [Krupa, D. (Ed.). (1997). Flight nursing core curriculum. Park Ridge, IL: National Flight Nurses Association. (pp. 752)].

15. b
The sympathetic nervous systems response to 'fight or flight' response includes increased respirations and heart rate, glyconeogenesis causing an increase in blood glucose, and the systolic pressure rises widening the pulse pressure. [Krupa, D. (Ed.). (1997). Flight nursing core curriculum. Park Ridge, IL: National Flight Nurses Association. (pp. 752)].

16. a
UHF 30 to 50 MHz is not a designated public service frequency. VHF low-band 30 to 50 MHZ is a public service system that follows the curvature of the earth and has the greatest range. VHF high-band 148 to 174 MHz is also a public service with less susceptibility to electrical disturbances. UHF 450 to 470 MHz is a public service and includes medical frequencies designated by the FCC. [Krupa, D. (Ed.). (1997). Flight nursing core curriculum. Park, Ridge, IL: National Flight Nurses Association. (pp. 761 - 762).]

17. d

The transport phase involves the transport of the patient to the designated destination, so taking report regarding the patient and determining the destination occur during this phase. Determining the nature of the call and obtaining detailed landing zone information occurs during the initial call or en route phase of a scene response. Determining the accepting physician's name is necessary for an interfacility transfer, but not a scene response. [Krupa, D. (Ed.). (1997). Flight nursing core curriculum. Park, Ridge, IL.: National Flight Nurses Association. (pp. 768).]

18. d

Air medical team members need to know the location of the American Embassy, be aware of any civil unrest that could jeopardize their safety during transport, and establish a plan should they become separated anytime while in the foreign country. Only pilot(s) need to be aware of altitude restrictions. [Krupa, D. (Ed.). (1997). Flight nursing core curriculum. Park, Ridge, IL: National Flight Nurses Association. (pp. 772).]

19. d

Fire poses a threat to the safety of the rescue team and must be contained before any type of extrication is attempted. [Krupa, D. (Ed.). (1997). Flight nursing core curriculum. Park Ridge, IL: National Flight Nurses Association. (pp. 777-778).]

20. b

Rescue begins by assessing and controlling hazards at the scene; scene safety is the priority for both rescue and health care providers. The next step is gaining access to the patient, wherever the patient may be, as well as maintaining access to communications. Once the scene is considered safe and the vehicle is stabilized, health care providers may attempt to begin urgent patient care. The next step is to enter the vehicle. The last four stages are disentangling the patient, packaging the patient for removal, removing the patient, and transporting the patient. [Krupa, D. (Ed.). (1997). Flight nursing core curriculum. Park Ridge, IL: National Flight Nurses Association. (pp. 779-780).]

21. d

All of the conditions listed, as well as an airway that cannot be secured, are reasons for immediate extrication of the patient. [Krupa, D. (Ed.). (1997). Flight nursing core curriculum. Park Ridge, IL: National Flight Nurses Association. (pp. 780).]

22. b

Having a plan in place is the most important step in dealing with any disaster. The plan should include the deployment of personnel and equipment. Practice drills prior to a disaster help educate personnel and evaluate strengths and weaknesses of the disaster plan. [Krupa, D. (Ed.). (1997). Flight nursing core curriculum. Park Ridge, IL: National Flight Nurses Association. (pp. 783).]

23. a

Medical Control in a disaster is performed by physicians. The transport team may be asked to perform patient care at the scene or enroute to the hospital, they may simply transport supplies and medical teams. [Krupa, D. (Ed.). (1997). Flight nursing core curriculum. Park Ridge, IL: National Flight Nurses Association. (pp. 784).]

24. a

By staging aircraft, resources can be monitored and deployed when necessary. Communication with the air medical team may also be more efficient. [Krupa, D. (Ed.). (1997). Flight nursing core curriculum. Park Ridge, IL: National Flight Nurses Association. (pp. 785).]

25. d
Disaster planning should be focused on the most likely type of disaster to occur in the area. Plans for a medium-size disaster are more likely to be used than those for a large-scale disaster. [Krupa, D. (Ed.). (1997). Flight nursing core curriculum. Park Ridge, IL: National Flight Nurses Association. (pp. 783).]

26. a
Emergent is not used. The four classifications most commonly used are "walking wounded" (no hospital treatment needed), non-urgent but hospital care needed, urgent-hospital care needed and dead or unsalvageable. [Krupa, D. (Ed.). (1997). Flight nursing core curriculum. Park Ridge, IL: National Flight Nurses Association. (pp. 786).]

27. b
The assessment portion of the teaching/learning process includes determining the strengths of the learner. Developing goals is a planning activity, assisting the learner to achieve stated goals is the implementation phase, and providing feedback is evaluation. [Krupa, D. (Ed.). (1997). Flight nursing core curriculum. Park Ridge, IL: National Flight Nurses Association. (pp. 794).]

28. a
The teaching/learning continuum is similar to the nursing process; such that identification of the barrier to learning is a priority. Once the barrier is identified (stress, inadequate initial demonstration, errors previously not identified, etc.), a course of action to correct mistakes in performance can be formulated. Although stress reduction techniques might be beneficial to overall learning enhancement, instituting these techniques alone, without identifying specific problems, will not assist a new nurse. In addition, the nurse may not need to see demonstrations of all equipment/procedures. Reassignment to another preceptor does not identify the learning deficits. [Krupa, D. (Ed.). (1997). Flight nursing core curriculum. Park Ridge, IL: National Flight Nurses Association. (pp. 798).]

29. d
Outreach education opportunities include presentations on landing zone safety, providing educational sessions to referring agencies, and interacting with community and political groups. The best community outreach, however, is for the flight crew to do the best job possible on each and every flight. [Krupa, D. (Ed.). (1997). Flight nursing core curriculum. Park Ridge, IL: National Flight Nurses Association. (pp. 806 - 807).]

30. d
Because children of different ages are at different developmental stages, the teaching/learning plan must be presented at a level that is understandable and interesting, based on the patient's age and current developmental stage. Children differ from adults in that they have a dependent personality. Motivation occurs for children through external rewards. Therefore, children must be able recognize a tangible reward for themselves for motivation to occur. Children's limited experience must be built on, as opposed to adults whose varied experiences may be used as a resource. [Krupa, D. (Ed.). (1997). Flight nursing core curriculum. Park Ridge, IL.: National Flight Nurses Association. (pp. 794).]

31. b
Flight nursing practice requires the recognition and implementation of a safe transport environment in collaboration with other air medical team members and those who are using the transport service. [Krupa, D. (Ed.). (1997). Flight nursing core curriculum. Park Ridge, IL: National Flight Nurses Association. (pp. 816).]

32. b

Standard V: Implementation the flight nursing core curriculum reads: "The flight nurse implements needed patient care interventions in an orderly, logical manner. [Krupa, D. (1997). (Ed.). Flight nursing core curriculum. Park Ridge, IL: National Flight Nurses Association. (pp. 816).]

33. c

The Standards are considered a framework or guidelines, not rules, regulations or laws. [Krupa, D. (Ed.). (1997). Flight nursing core curriculum. Park Ridge, IL: National Flight Nurses Association. (pp. 819).]

34. b

The referring, or transferring physician or provider must document clearly the medical necessity for transfer. [Krupa, D. (Ed.). (1997). Flight nursing core curriculum. Park Ridge, IL: National Flight Nurses Association. (pp. 821).]

35. a

Indemnification occurs when an employee is found professionally negligent and through vicarious liability the employer has to pay the settlement or fine. The employer could then require the employee to pay back the money. [Krupa, D. (Ed.). (1997). Flight nursing core curriculum. Park Ridge, IL: National Flight Nurses Association. (pp. 822-823).]

36. d

Ethical guidelines determined by the Department of Health and Human Services include the right to anonymity, privacy, and self-determination. [Krupa, D. (Ed). (1997). Flight nursing core curriculum. Park Ridge, IL: National Flight Nurses Association. (p. 831).]

37. c

The dependent variable in an experiment is the variable that the investigator studies to determine the effect of the independent variable. An independent variable is the variable that is systematically manipulated by the investigator. [Krupa, D. (Ed.). (1997). Flight nursing core curriculum. Park, Ridge, IL: National Flight Nurses Association. (p. 830).]

38. d

Quantitative research is the traditional format in nursing studies. Descriptive, experimental, correlational and quasi-experimental re all types of quantitative research. [Krupa, D. (Ed.). (1997). Flight nursing core curriculum. Park Ridge, IL: National Flight Nurses Association. (p. 833).]

Answer Page to blank spaces in the book

Page 17

As altitude ↑, The gas **expands**

As altitude ↓, the gas **contracts**

Effects on patients

a.　　**Barotitus Media**

c.　　**Barodontalgia**

d.　　**Barosinusitis**

e.　　**Barobariatrauma**

Effects on equipment

a.　　**ETT cuffs**

b.　　**MAST trousers** (If you got em')

c.　　**Gravity fed IV drip rates**

Gas volume expands as temp **increases**

Gas volume shrinks as temp **decreases**

Relationship between temp and **pressure**

Page 18

**Drugs, Exhaustion, Alcohol, Tobacco,
HYPOglycemia**

Page 19

Obstruction of the **RCA**

Infarcts the **Papillary Muscles**

Obstruction of the **LAD** aka the "widow maker"

Obstruction of the **Circumflex**

Page 20

Elevation in leads **V5** and **V6**

Page 22

V2-V4	**Anterior**
II, III, aVF, V5, V6	**Inferior**
V1-V3	**Anteroseptal**
I, aVL, V4-V6	**Lateral**
V4R – V6R	**Right**
V1, V2, ST Depression	**Posterior**

12 Lead is a **Anterio-lateral wall MI**

Page 24

What vessel? **Right Circumflex Artery (RCA)**

Page 31

MI
Acidosis　　　Resp and Metabolic
Tension Pneumothorax
Cardiac Tamponade
Hypoxia, Hypovolemia, Hypothermia,
Hypoglycemia, Hypo / Hyperkalemia
Embolus (Pulmonary)
Drug Overdose

Page 38

↓	↑	
↑	↓	
↓	↑	HCO3 off too though
↑	↓	HCO3 off too

Page 42

ELT @ **4 G's**

Frequency @ **121.5**

Page 42 (continued)

After **15** minutes

Takeoff, Taxi, and **Landing**
Eyes **Outside**

Notify **Pilot in Command (PIC)**

During **Takeoff**

Min. of **4**

Rating of **2000** hours

Page 43

VFR Weather Minimums per <u>CAMTS</u>

Local Day   **500 ft ceiling, 1 mile vis**

Local Night   **800 / 2**

Cross County Day   **800/2**

Cross Country Night   **1000/3**

VMC- **Visual Meteorological <u>Conditions</u>**

I-IMC- **Inadvertent - Instrument Meteorological Conditions** (KILLS YOU)

VFR – **Visual Flight <u>Rules</u>**

IFR –   **Instrument Flight <u>Rules</u>**
       Anything less than VFR min's

During an Emergency...
1. **Lay patient flat**
2. **Assure everyone strapped in**
3. **Turn off O2**
4. **Secure equipment**
5. **Confirm your seatbelts**
6. **Visors down**
7. **Assume crash position depending, on direction your facing**

Post Crash Procedures
Throttle, Fuel, Battery, Exit, Help crew if able, Meet at 12:00 nose
ELT activation **4 G's**

Frequency? **121.5mHz**
(new 406 effective 2/09)

Exit and meet at **12:00 nose**

Secure

A.   **Shelter**

B.   **Fire**

C.   **Water**

D.   **Food**

Rule of 2's (means you can live...)

In 2 hrs of **extreme cold**

For 2 days **without water**

For 2 weeks **without food**

Page 44

RW weather minimums

FAA says **"See and avoid clouds, at an altitude that will ensure a safe landing in case of an emergency**

For CAMTS, **See page 43 Answers**

We meet at **12:00 on the nose of the A/C**

PAIP activated within **15** min...

Undergarments ¼"

First priority is **Shelter**

Required by CAMTS?
Helmets   **YES** if RW, **NO** if FW
Nomex   **No**, but flame retardant is
Own dispatch **No**, look at REACH, they
                  dispatch for several programs

Eyes where... **Outside**

Voluntary guidelines?
CAMTS   **YES**
AAMS Weather mins **YES**
FAA   **NO**
OSHA   **NO**
NTSB   **YES**, unless it is
                       damage > $25,000 or
                       serious bodily injury

Everyone is responsible for safety

Sterile cockpit is **Silence during Taxi, takeoff, and landings**, unless it deals with an emergency at hand.

Lack of will to survive is #1 killer of people in survival

Inflate flotation devices **when you exit the aircraft** (duh!)

In water, you exit the a/c **when all motion stops**

Head Injuries are the most common cause of death from injury following an accident

Emergency freq is **121.5**

Shelter best location? **This is a class discussion, too many variables to list**

The most important characteristic of survival is to **not give up**

Page 45

No, we need to maintain a "sterile cockpit" as this is a <u>critical phase of flight</u>.

Page 47

Hypoxic Hypoxia is a **deficiency in alveolar oxygen exchange**

Hypemic Hypoxia is a **reduction in the oxygen carrying capacity of the blood**

Stagnant Hypoxia is **Pooling of blood, restriction of blood flow, or a reduction of tissue perfusion.**

Histotoxic Hypoxia is **the impaired cellular ability to use oxygen**

Barotitis Media - **Middle ear disturbances**

Barosinusitis - **Sinus membranes**

Barodentalgia – **Dental cavities**
Respiratory System – **Previous Pneumothorax, Chest drainage systems**

Medical equipment – **ETT, IV containers, MAST trousers, Pneumatic ventilators**

Page 48

Signs of dehydration
    Dry mucous membranes
    Sore throat, Thirst
    Scratchy eyes

Flicker Effect
    The sunlight "flickering" through the rotor blades while the patient is facing the light, can cause a seizure if predisposed. (Pre-eclampsia)

D - **Drugs**
E - **Exhaustion**
A - **Alcohol**
T - **Tobacco usage**
H - **<u>HYPO</u>glycemia**

Page 69

History:
a. **Chief Complaint**
b. **History of present illness**
c. **Name and Amount of substance**
d. **Concurrent ingestion**
e. **Time of exposure**
f. **Method of exposure**
g. **Circumstances of ingestion**

Page 83

Brudzinski's	**Neck**	
	Meningitis	
Kernig's	**Hamstring**	
	Meningitis	
Cullen's	**Bruise & Periumbilical pain**	
	Pancreatitis	
Grey Turner's	**Flank & Groin bruising**	
	Pancreatitis	
Kehr's	**Referred Left shoulder pain**	
	Splenic	

Murphy's	**Palpate RUQ & breathe in Gallbladder**
Levine's	**Clenched fist over chest Cardiac**

Page 90

Treatment   Since they are usually self limited seizures, we will address
- **Stabilize ABC's**
- **IV Access**
- **Mag Sulfate IV 4 Gms over 20/min** Antidote for Mag toxicity is Calcium Gluconate 1 Gram
- **Dilantin 100 mg IV**
- **Benzo's (Valium most likely)**

HELLP Syndrome
Hemolysis & Hypertension
Elevated Liver Enzymes
Low Platelets

PIH **Pregnancy Induced Hypertension**

The "Big Three" include...
- **Pulmonary edema**
- **D.I.C.**
- **Renal failure**
- **Hepatic failure - liver rupture**
- **H.E.L.L.P. Syndrome**

Page 98

Transducer at the **Phlebostatic Arch**, which is at the Midaxilliary line, at the level of the third rib when patient is supine

Page 107

Pt with a temp of 102:		**R**
Pt with a CO2 of 30:	**L**	
Pt with ↓ 2,3,DPG:	**L**	
Hemophiliac:		**R**
pH of 7.50:	**L**	

CO2 of 50:		**R**
Frozen lake drowning:		**Trick question**

Can be either, as ↓ temp is a left shift, however acidosis (Resp or Metabolic) is a right shift (depends if "dry" drowning)

Ph. of 7.20		**R**

Page 138

Fx commonly assoc with compartment syndrome is the **Tibia**.

Signs of compartment syndrome are **Pallor, Pulselessness, &Sensory Loss.**

Nursing intervention for open wound include **covering with a moist sterile NS dressing**.

Inflammation of the synovial cavity surrounding a joint is called **Bursititus**

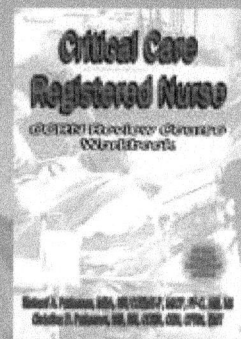

www.ingramcontent.com/pod-product-compliance
Lightning Source LLC
Chambersburg PA
CBHW061354210326

41598CB00035B/5976